Martial Culture in the Lifeways of US Servicemembers and Veterans

This book develops a new concept—"martial culture"—with which to problematize and reframe thinking surrounding the lifeways of US servicemembers, by exploring the values, beliefs, norms, and rituals they are exposed to and practice during military service.

By reuniting the two concepts of servicemember and veteran into one overarching cultural model, the author shows how the concept of *martial culture* can be used to acknowledge the unbroken, holistic, multidimensional life cycle of an individual. Adopting a comparative mythological approach and drawing upon Roman, Navajo, Hindu, Norse, and Japanese myths that speak to the lived experiences of servicemembers, veterans, and their families, it weaves together ancient voices and contemporary servicemember experiential existences to offer new insight into the psychological experience of servicemembers.

It will be of strong interest to psychologists who seek to develop their treatment of veterans by understanding the unique lifeway of service without judgement and offering a balanced, integrated spiritual connection, while pushing back against both inaccurate assumptions of martial lifeways and the influences of industrialized secular approaches to service. It will also appeal to those within the fields of military sociology and psychology.

Nathan J. Hogan obtained his PhD in Mythological Studies with emphasis in Depth Psychology from Pacifica Graduate Institute, USA. He served 22 years of active service in the US Air Force and Army, and his research interests include mythology, history, and the intersection of human belief and behavior.

Routledge Studies in Leadership, Work and Organizational Psychology

Martial Culture in the Lifeways of US Servicemembers and Veterans

Military Psychology, Ancient Mythology, and Re-Souling Service

Nathan J. Hogan

Routledge
Taylor & Francis Group

NEW YORK AND LONDON

First published 2024
by Routledge
605 Third Avenue, New York, NY 10158

and by Routledge
4 Park Square, Milton Park, Abingdon, Oxon, OX14 4RN

Routledge is an imprint of the Taylor & Francis Group, an informa business

© 2024 Nathan J. Hogan

Library of Congress Cataloging-in-Publication Data
Names: Hogan, Nathan J., author.
Title: Martial culture in the lifeways of U.S. servicemembers and veterans: military psychology, ancient mythology, and re-souling service / Nathan J. Hogan.
Other titles: Military psychology, ancient mythology, and re-souling service
Description: New York, NY : Routledge, 2024. | Series: Routledge studies in leadership, work and organizational psychology | Includes bibliographical references and index.
Identifiers: LCCN 2023042200 (print) | LCCN 2023042201 (ebook) |
ISBN 9781032601366 (hbk) | ISBN 9781032612638 (pbk) |
ISBN 9781032613222 (ebk)
Subjects: LCSH: Psychology, Military. | War—Mythology. | Sociology, Military—United States. | Soldiers—United States—Psychology. | Veterans—United States—Psychology.
Classification: LCC U22.3 .H64 2024 (print) |
LCC U22.3 (ebook) | DDC 306.2/70973—dc23/eng/20231117
LC record available at https://lccn.loc.gov/2023042200
LC ebook record available at https://lccn.loc.gov/2023042201

ISBN: 978-1-032-60136-6 (hbk)
ISBN: 978-1-032-61263-8 (pbk)
ISBN: 978-1-032-61322-2 (ebk)

DOI: 10.4324/9781032613222

Typeset in Times New Roman
by codeMantra

Contents

Preface

This work is the culmination of numerous experiences that intertwined to inspire the creation of this book as an attempt to introduce mythologies to demonstrate the rich, multidimensional, and holistic human culture of the servicemember and veteran as well as challenge two-dimensional stereotypes prevalent in professional, academic, and popular discourses concerning those who serve in the military. This began with my early interest in and eventual entrance into service and my love of different cultures and their mythologies. However, it was really a series of experiences that occurred as I began to leave the military that culminated in the concept of martial culture, highlighting the cultural differentiations between martial and civilian cultures.

The first experience was my introduction to the University of Arizona's VETS (Veterans Education and Transition Services) Center. Like many servicemembers, I had spent my military life so immersed in the webs of my daily life that outside concerns rarely made their way to my consciousness. However, I had heard of a then-recent trend of providing safe spaces for students on college campuses. The necessity of this practice sounded strange to those of us who had routinely volunteered for difficult assignments and combat deployments. Yet as I stood in the VETS Center, facing my own impending expulsion from the military, I wondered, "Are veterans a special population?" My interactions with student veterans convinced me that safe space had less to do with physical threats but the necessity of peer support where language and behaviorisms that felt comfortable could be expressed openly.

The pivotal event that sparked my idea of cultural differences between civilians and military was my mandated participation in the TAP program. Congress ordered this program to address the issues of veteran joblessness and homelessness. The goal was to create a program that introduced soon-to-be veterans to employment, education, medical, and other opportunities. It covered everything from resume writing to dressing for an interview.

While the classes covered things veterans should do when assimilating "back" into civilian society, there were also informal nuances of things veterans should not do. These kept creeping into the conversation: how not to dress, what not to say, avoid military vernacular, and so on. I was struck at the similarities between what I was hearing and cultural briefings we would receive before visiting foreign countries. These are commonly referred to as cross-cultural briefings and are designed to sensitize servicemembers to the values, beliefs, and norms of behavior of the foreign culture(s) in which they will find themselves. The briefings help to avoid social faux pas and recognize cultural cues. Knowingly or not, the TAPS program recognized that servicemembers leaving the military organization were going to be sent into another foreign culture. Looking back at that moment, it was the first time I realized I was becoming an emigrant.

Further, I had an informal conversation with a fellow student in my graduate program for mythology. I was explaining my germinating thinking regarding the holistic culture of servicemembers. Her response was to describe her life experience growing up in a commune where all things were shared and everyone had a hand in the daily life cycle and her work later in life for a county government. She did not think that the county government job contained the necessary characteristics required for a holistic culture and therefore, the military, which she considered the same as civil service, would not have those deep familial connections. I was surprised by the comparison she drew between government civil service and life in the military. My response was simple. Of the two, being a servicemember is far more akin to her description of the commune. It was that interaction that solidified my interest in creating the theory of martial culture.

Lastly, throughout my academic journey, I have noticed that balanced martial cultural voices, especially concerning mythology, are overpowered by traditional academia originating from civilian experiential thoughts. Many academics who engage with veterans foray into a world strange to them. They are outside the culture, and lack the cultural competence to understand what they observe. When these people return to the shores of their institutions, their reported results are heavily biased by the impetus that sent them to seek out servicemembers, veterans, and their families in the first place.

The clamor of professionals and academics outside of martial culture who have the greatest influence in discussions about martial cultural peoples asks whose voice is truly being heard. I have attended several lectures by academic specialists from various fields (neuroscience, classical literature, psychology, and religious studies) speaking at conferences about their participation in projects with veterans, servicemembers, or their families. Often, they are viewed by other academics as experts on martial culture. Their discussions focus on a very narrow experience of martial lifeways—usually war and homecoming as tragedy—and the weight of the discussion is usually tipped toward the implicit bias of the academic in how they view war, veterans, servicemembers, or their families. Without much divergence, most approach veterans with their specialty in a salvific or missionizing fashion in order to correct, save, or fix veterans, without the grounding of cultural knowledge.

Despite their cultural blindness, these individuals, with no experience of the martial, have outsized voices in both academia and government organizations such as the US Veterans Administration. There are an overwhelming number of academics, practicing psychologists, and therapists who have no connection to martial culture except through media, social media, and/or students or patients. In one conversation, I was informed that my experience and study did not outweigh what an individual had read throughout his career. In the most extreme cases, academics will quote other academics about war when neither have ever experienced war. Veteran-related events or messaging that supports the views of these individuals are promoted and any events or messages that conflict are summarily dismissed.

This book seeks to address the holistic aspects of martial life through mythology. While some might view this attempt as romanticizing conflict and warriorhood, nowhere in this book is there to be found a love of war. It focuses on those who live lives dedicated to service, attempts to expand the discussions surrounding servicemembers, veterans, and their families and challenge stereotypes. It also seeks to speak to those people within martial culture. My experiences within martial culture demonstrate that in many ways, emphasis on industrial and secular aspects of the military organization inhibits the holistic human experience of service. I hope that a reinfusion of mythological connections will awaken current servicemembers, veterans, and their families to their spiritual inheritance of these myths.

Acknowledgments

I am particularly grateful to Professor Patrick Mahaffey for his infinite well-spring of support and encouragement in pursuing this project. My humble gratitude to Dr. Kris Alstatt for the guidance he gave me in finding my way through the literature of the anthropology. The misunderstandings and misinterpretations that resulted are my own. My undying appreciation to my dear friends and editors, Dr. Kath Sargent and Dr. Dennis C. Hall, who worked tirelessly to help in the creation of a much finer work than I could have ever dreamed of. I owe an immeasurable debt of gratitude to the servicemembers and veterans I have served with and known and unfortunately are too numerous to mention here. Not a day goes by that I do not think of the soldiers, sailors, airmen, marines, and coastguardsmen I served with, to include those belonging to foreign militaries. Much of the inspiration for the concepts in this work came from those connections. My thanks also to the veterans that trusted me enough to agree to allow me to interview them about their experiences. I would like to note the support I received from my friends and colleagues, particularly Mr. David Nelson-Fischer, Mr. Zach Blackwell, Mr. Dusty Miller, and our entire team who allowed me to work through ideas and concepts that appear in this book, providing unvarnished and invaluable feedback. I am grateful to my family for their support and encouragement. Finally, my eternal gratitude to Rachel for her patience in accepting long hours of research and writing and keeping me encouraged to finish this immense project.

1 Introduction

This book is written for those who are interested in an exploration of martial cultures as described in mythologies and how those myths may have contributed to, and have correlations with, the current US martial culture as it exists today. *Martial Culture* speaks to two broad audiences. The first is those who have not had any engagement with this culture other than what is propagated through social media, television, film, and other forms of media, but wish to engage with servicemembers, veterans, and their families from various perspectives of health care, vocation, and social engagement. This book is also for servicemembers, veterans, and their families whose lifeways make up what I refer to as US martial culture.

For the first group, the exploration is a way of knowing people who live very different lives than most others, perhaps much more similar lives with other martial cultures than those within civilian culture. For the latter, this foray offers a way of reflecting on the various holistic ways in which martial cultures in the past have crafted their experiences into mythologies, thereby providing a reacquaintance with familiar foundational values, beliefs, and norms of behavior found in the mythologies that underpin lives that experience a full range of human experience.

The individual and collective experiences, ritual participation, and symbolism define and describe the ways a martial lifeway shapes the individual's interactions with themselves and the external world. Mythology is a way for a person or group of people to describe their sociological, pedagogical, psychological, and spiritual world experiences. I use an interdisciplinary approach that utilizes anthropology, psychology, history, and spirituality studies to define martial culture and explore diverse mythologies through a martial culture lens to glean the ways they can provide guidance to those who serve today.

As a mythologist, I have found that those who serve in the US military, as well as their families, live in a world of myth and ritual that form the webs of their cultural identity. They experience beautiful, amazing, rich, deep, and sometimes heartbreaking rituals and symbolism nearly every moment. Some are aware of this, but most are not. For the servicemembers and their families who

DOI: 10.4324/9781032613222-1

are unaware of this mythical foundation and spiritual support, the experience of many servicemembers and their families is one of existential crisis. Because the cultural rituals and symbols are largely disconnected from the roots that filled them with soul and meaning, modern secular militaries have lost that internal source of support. The extreme cases of those crises result in psychological and/ or physical breakdowns, while the more common version is simply a hollow feeling of disconnection from something that defies description. The accompanying sense of aloneness is exacerbated by the inevitable ejection, through separation or retirement, of the military servicemembers out of their culture and into civilian life. The damage inflicted by being excommunicated, desired or otherwise, is compounded by the culture shock of being surrounded by strangers who prefer that servicemembers "leave it behind and get on with life" despite the aching sense of echoing loneliness (Guthrie-Gower and Wilson-Menzfeld; Smith; Hannel).

US martial culture is beset by various issues ranging from psychological diagnoses, physical injuries, toxicity, careerism, sexual assault, racism, military family crises, substance abuse, and suicide. Legions of health professionals and military leaders have worked for decades to attempt to treat the symptoms and unravel the issues experienced by a military modeled on secular industrialism. That is, a prescriptive model of civilian-as-citizen mobilized at the behest of the government to become a servicemember and then returning to a civilian state when a demilitarization is determined necessary by that same government. Embedded within this applied concept is the assumption a person can compartmentalize two quite separate cultural identities that conflict in key values. This practice by the government creates a vacillation between two quite different states of being, requiring values, beliefs, and norms of behavior quite different from each other, which has caused misunderstanding, confusion, and crisis.

The US military has transformed from a combined conscript/volunteer force to an all-volunteer organization. This transformation was accompanied by concerns that the US people and their military would slowly grow apart. Martial culture, with its accompanying acculturative practices, already creates a cultural differentiation between martial and civilian cultures. The difference in the US military is that for the last 50 years a person has been given a choice as to whether to enter martial service or continue in civilian culture.

Thousands of generations of humans have been inducted into cultures of martial service and have trod familiar ground, left behind guideposts in the form of myths. The problem, stated bluntly, is they are no longer remembered, much less heard. A number of myths were written by those who engaged in martial service. It is time to reconnect those myths with those who serve in military cultures today.

Without words to describe the emotional or spiritual effects, modern military members still feel their connection to ancient wisdom. The military family does not speak of their world as containing myth or ritual; they live it unconsciously.

The culture in which they live is not described in terms a religious scholar, anthropologist, sociologist, or psychologist would use. Reducing every gesture, spoken word, or symbol down to some sort of artificiality erases the essence of what it was to be within this all-pervading *kosmos*.[1] They live in a world at least a half-step removed from what many would refer to as the *normal* world, the term used to describe civilian society.

This book endeavors to accomplish two things. The first is to introduce a theory of martial culture to reframe the dialogues and challenge existing paradigms surrounding how veterans and servicemembers relate to both the external environments around them and the internal worlds in which they live. It should be noted that veteran and servicemember are gender neutral terms. Past models in psychology that endeavor to apply qualities associated with martial culture to gender, dumped into masculine and feminine buckets, no longer apply, and never should have been attempted. If anything those typologies did more damage by endorsing sexism and male dominance in military organizations. When people have resorted to force, no one who showed aptitude for martial service was exempt.

The outdated arguments of male crises or masculine crises as synonymous with military service and aggression need to be brought into accord with the reality that to participate actively in martial culture is a holistic human experience. Studies in women veterans issues and men veterans issues are subordinate to what I refer to as martial culture, which is the holistic totality that includes servicemembers and veterans' families. The Orphic Hymn to the Goddess Athena notes that gender is not a determinant for inclusion in martial culture, "O, warlike Pallas, [...] Female and male,[2] the arts of war are thine" ("Orphic Hymn to the Goddess Athena").

Further, though largely outside the scope of this book, martial culture is inclusive of the animals who have both trained and served in martial capacities. Two ready examples are the partnerships of horse and dog with servicemembers throughout time in martial culture (the Hindu Mahabharata gives excellent examples of this). This inclusion would add the disciplines of animal behaviorism and psychology as subsets to be included in martial culture. The study of the theory of martial culture must be broad enough to take in all experiences.

Much of the writing on why veterans feel disconnected has been focused on the experience of combat. I argue that the holistic embodied changes, physiological as well as psychological, begin from the moment the decision is made to enter martial culture and the experiential aspects that shape the "bottom up" transformations at the moment basic training begins. These changes are holistic, signifying that the individual has crossed from civilian culture to martial culture through an expedited acculturation process.

Parallel research in the past few years has produced studies that corroborate the holistic and valid acculturation servicemembers, veterans, and their families' experience (Shepherd et al.; McCaslin et al.; Smith; Hannel). I argue martial acculturation must be recognized and accepted as a natural event of growth in a different culture rather than demonizing it as a form of traumatization or indoctrination, which, in turn, creates greater schisms both inside the veteran/servicemember cultural sphere and between civilian and martial cultures. Further, the veteran must be viewed as an immigrant of another culture and we should embrace cultural pluralism as a way of bringing their culture into the tapestry of other cultures within the United States and the greater world community.

The second goal of this book is to explore four mythological traditions that address martial service and determine the parallels to contemporary servicemember and veteran experiences. It explores the similarities between different martial cultures separated by time and space examining how their mythologies determined the best way to sustain a martial lifeway while continuing to live forward, incorporating the beautiful aspects of service while weathering the harsh difficulties encountered while serving.

Culture

Military culture is almost exclusively described by sociologists as a secular organizational culture. A typical description as offered by Allan English states that military culture consists of three spheres, "the communal character of military life, the heavy emphasis on hierarchy, and military discipline and control" (39). While these three aspects hold true for military culture, the power of the organization to enforce hierarchical models and "discipline and control" is left behind when the servicemember becomes a veteran. English's latter two aspects frame the discourse of the study of military culture in terms of organizational culture, similar to that of a large company or governing agency such as Google or the US House of Representatives. Yet the first aspect, the "communal character of military life" is, I argue, not a character at all, but a real communal, holistic culture. Within this community, there are values, beliefs, and norms of behavior that, through the processes of acculturation, ritual trials, and formations of deep and enduring relationships, will be embodied within the individual to such an extent that they endure well after one leaves military service. It is this linear development of the individual as part of a culture that I assert requires validation.

In *A Secular Age*, Charles Taylor divides the experience of the individual into private and public spheres and argues it is the private sphere the private versus the public sphere in which the private sphere nourishes the human. Within US martial culture, the individual begins with no actual private sphere and, in many cases, builds a lifeway filled with a strong web of relationships that weave the "private" and "public" aspects of martial life and military service together so tightly there is almost no way to distinguish one from the other. This does not consider the very structure of the military, where the regulations allow the organization far more latitude to both access and control the individual's private life.[3] In practice, there is little in the way of separation between public and private, and therefore a greater need for deemphasizing secularism to the extent it detracts from the communal and transcendent qualities of martial life.

By secularism, I mean the public aspects of military duties that typically consist of the career elements inherent to belonging to the military organization. Examples include evaluations of skills, annual evaluation or fitness reports, studying for qualifications, promotion testing, and supervisory duties which are typically found in other government and business organizations. One of the interesting commonalities of this list is the focus on the individual's performance as a singular entity rather than as a member of a group, team, or tribe. Secularism, in application, sterilizes the transcendent. To be thought of as a government employee confuses both those within martial service and those whom they serve as to what is the place and role of military culture within the cosmos. As some servicemembers note, there is a point at which all interactions become transactional rather than genuine connections (France).

From an outsider perspective, this secularly informed objectivity has dominated the framing of the military today. Despite the nearly constant presence of ritual or symbolism, the idea of a secular military has, to a degree, stolen the soul out of values and beliefs. What were once sacred rites and symbols are now traditions and customs. The deep-rooted feeling of connection to an ancient self as part of a tribal group is still felt but lacks vehicles of expression, while the emotional and spiritual experiences are explained away as a biological and psychological phenomenon, and potentially discouraged, even from within. Today the word military is equated to a rational, stoic mind and body. This mind and body suffer, I argue, from a split from soul.

Within certain circles of the military, there is both awareness and active participation in martial rituals. People within those circles are conscious of the *participation mystique*, or mystical participation. Yet, many others simply go through the motions of the utilizing symbols, rituals, and myths. With the severing of their connections to their mythic past, these symbols, rituals, and myths have become dried out, misunderstood relics. While some acknowledge this mystical participation, it is largely ignored by academics discussing military culture. The term *martial* reinfuses mythic inheritance with military culture. It implies a holistic unity filled with human customs, rooted in mythological story that is reinforced with ritual and symbolism, and is brimming with a multiplicity of human archetypes. In short, the *martial* is the soul that must be much more fully integrated into the rational body of the military and extends beyond membership in a military organization.

Approaching a study of martial culture from a secular, industrial organizational context can neither resolve nor provide answers for some of the most fundamental questions and conflicts a human can face—not in theoretical rhetoric—but in concrete forms where mud, sand, heat, cold, utter exhaustion, and deprivation meet flesh, blood, bone, mind, and soul.

This book demonstrates the existence of martial culture; elucidates the myths, symbols, and rituals associated with it across four regional cultures; and reconnects contemporary US martial culture with its mythic origins by exploring mythologies and how they correspond with US martial culture in the areas of the physical, psychological, sociological, and spiritual.

Further, this book seeks to illuminate the underlying themes that are repeated irrespective of time, region, and people. These themes inform the great myths which relay a transcendent reality that connect martial cultures across space and time. It is this reality that must be reconnected to current military cultures who have been cut off from their myths. To do this, it is necessary to introduce the myths to those who have feelings and emotions that cannot be expressed because the roots of the language needed to elucidate those feelings have been largely forgotten. By reintroducing the stories of the forebearers, this book may inspire those who need to hear the ancient wisdom and gain grounding and strength from an understanding that they exist as inheritors of service.

The myths chosen were, to the extent possible, from cultures that are not necessarily European. The question I posed was, "What do non-Mediterranean based myths have to say about martial service?" My theory is that there might be different philosophical thinking to under-gird the lifeway of a warrior. Additionally, typical engagements with veterans and myths are largely focused on Greek and Roman mythologies. For this reason, I have consciously chosen to not engage either Roman or Greek mythic traditions in an effort to explore potentially different relational models to martial cultural norms.

Mythology

Myth, mythology, and mythos are terms usually regarded negatively in the modern world. As Patrick Mahaffey notes, myth "is often regarded to be an untruth [....] Ancient Greek philosophers distinguished mythos from logos: from mythos came intuitive narrations; from logos reasonable explanations" (25). A mythology is a sacred story or collection of stories that expresses "subjective truths about the mystery of life that cannot be articulated in ordinary language" and is "sacred to and shared by a group of people wherein they find their most important meanings" (Mahaffey 25). Myths are complex and serve as the basis from which various disciplines of knowing and engaging in the world and in the self emerge from. As a mythologist, I approach each religious text or oral tradition with the reverence and respect of the transcendental nature intended.

According to Joseph Campbell, mythologies function to support the individual through the four functions of the mystical, the cosmological, sociological, and psychological. Within US martial culture, there is a lack of recognition of the transcendent nature of service and in many cases a failure to provide a space or language for the individual to inculcate and integrate the mystical experiences inherent in martial service.

The third function, the sociological aspect, creates "the moral, social sphere [...providing] the validation and maintenance of an established order" (Campbell *Creative* 621). If anything sets US martial culture apart from the various civilian cultures of America, it is this function. While usually described in a secular way as customs, the rituals that maintain the social order of martial culture are unique. Further, the codified moral order, the rules and regulations which govern the US military, only applies to martial culture.

The fourth function of myth, what Campbell refers to as the psychological aspect, serves "the centering and harmonization of the individual" (Campbell *Creative* 623), which provides balance and completeness within the other three dimensions of the living mythological realm in which the individual exists. To a certain extent, US martial culture does follow Campbell's assertation that in traditional societies the individual is meant to give "of oneself, and even giving up of oneself altogether" (Campbell *Creative* 623).

The ancient mythologies explored in the following chapters demonstrate various mythos which serve the individual who participates within martial cultures. Campbell's four functions of a living, valid mythology serve as a basis for how each myth is explored in relation to US martial culture. Mythologies serve as both a representative of past peoples who have served martial lifeways and potential guides for how to live a martial lifeway.

Introducing Martial Culture

Most people today see the military as a subculture encompassed by the larger civilian culture; that is, while separate, they are not that dissimilar. There are fundamental principles within both the civilian and military cultures that result in differences in worldview and approaches to lifeways. Those differences demonstrate that they are actually two different cultural groups, not a culture and a subculture.

Therefore, martial culture must be both defined and described before beginning an exploration of mythologies concerned with the martial and how they are relevant to modern members of the military. The word *martial* in this work seeks to reintroduce and reintegrate the multidimensionality of human existence within and after military service through a useful revivification of mythology.

Culture

While those who use the term military culture view the military as an organization or institution which simply represents a slice of civilian culture in microcosm, martial cultures are separate cultures from civilian cultures and not a sub-culture of the latter. The etymological roots of the term *culture* refer to cultivation of land, by the eighteenth century it had evolved in Germany to refer to ideas, customs, and behavioral norms of a group within a society. The line between cultures is different than that of a culture and its sub-cultures. This is significant because martial culture is a separate web of metaphors the service-member, veteran, and family members interact with. These interactions are both external and internal, forming a lifeway of service.

There are various definitions and descriptions about what culture is abound. While Clyde Kluckhohn had several definitions for culture,[4] anthropologist Clifford Geertz follows Max Weber's idea "that man is an animal suspended in webs of significance he himself has spun, I take culture to be those webs, and the analysis of it to be therefore not an experimental science in search of law but an interpretive one in search of meaning" (5). Geertz further explains that "As inter-worked systems of construable signs (what, ignoring provincial usages, I would call symbols), culture is not a power, something to which social events, behaviors, institutions, or processes can be causally attributed; it is a context" (14). Context is vital for understanding martial cultural symbols and rituals. From an ethnographic perspective, without the surrounding cultural context, interpreting the trappings of martial cultures is extremely difficult.

There are yet other examples of what culture may be, "'A society's culture,' to quote Goodenough [...] 'consists of whatever it is one has to know or believe in order to operate in a manner acceptable to its members.' [which allows for a] writing out of systematic rules, an ethnographic algorithm, which, if followed, would make it possible so to operate, to pass (physical appearance aside) for a native" (11). This, according to Geertz, is the central understanding for those who are proponents of "ethnoscience, componential analysis, or cognitive anthropology" (11). In this view, the more ritualistic and guided the cultural system, the more difficult it would be for one to pass as a member of that system. The US military has many volumes that cover acceptable rules of conduct within their martial culture, with numerous addendums addressing variances within services and even units. With this as a qualification for differentiating between one culture and another, the highly ritualistic and guided US martial culture would prove difficult to distill into an ethnographic algorithm that would allow for someone outside of it to pass for a native.

The US martial culture of today is influenced by and in turn influences the civilian cultures within the United States; influence should not be confused with likeness or sameness. Many cultures have existed next to each other and closely influenced each other while each remaining both unique and distinct from the other. As David Graeber and David Wengrow assert, cultures may be created through a "process by which neighbouring groups began defining themselves against each other and, typically, exaggerating their differences" (504). This phenomenon suggests that US martial and civilian cultures may actually exaggerate their differences to enhance difference, I believe to the detriment of both. One way to reduce this exaggeration is to formally acknowledge both cultures as valid lifeways, each with its own unique beauty.

Geertz, in discussing what anthropological interpretation is and how it should be approached, explains "our formulations of other peoples' symbol systems must be actor-oriented" and the cultural constructions should be driven by "the formulae they use to define what happens to them" (14). An exploration of martial culture must be focused on the normal everyday experiences of its members. This normal is defined by the commonalities found in the experiences of the cultural group. "Understanding a people's culture exposes their normalness without reducing their particularity [...] it renders them accessible: setting them in the frame of their own banalities, it dissolves their opacity" (Geertz 14). If one does not "expose their normalness without reducing their particularity" thus allowing them to be understood, one can easily become hostile to certain behaviors, symbols, rituals, or any other act which is, in one's own culture, aberrant. Ignoring cultural context and imposing one's own cultural norms can lead at best to misunderstanding, at worst to pathologizing and demonizing.

Civil

Martial culture as discussed in this book exists only in relation to what differentiates it from civil[5] or civilian culture. I use the term civilian here to describe those persons who have gone from birth to death without having experienced military or martial existence. This offshoot of a certain type of human experience has existed for thousands of years and has steadily increased in number. The rise of paid professional military organizations, as well as the rise in an increasingly diverse number of roles within civilizations, exponentially increased the number of persons within a society who have not been exposed to military service, either by happenstance or by choice. By the time the term *civilis* came into use, the status of a civilian as a separate existence from a warrior/soldier was already deeply entrenched in human experience.[6]

The introduction of the person who could move from birth to death without participation in organized force—defined as activities such as a group hunt, defending against raiding parties, or an organized military force, etc.—took shape and grew to embody the idea of civil and civilization as a lifeway. Civilians are the cultural inheritors of this ordering principle.

Martial

Martial[7] was originally intended to mean the "arts of Mars." This term encapsulated all aspects of martial lifeway as well as proficiency in all skills associated with that lifeway. Today it has largely been used to describe various arts of unarmed styles of combat or health practices. In psychological and mythological fields, the root of martial, the Roman deity Mars, is most often caricatured as a bellicose, perhaps cruel, warrior. However, the Romans took a far more nuanced understanding of life in war and peace and the necessity of the entire pantheon to be included, and even Mars had multiple "faces."

Contemporary use of the mythology has reduced Mars to a singular domain and function. Yet, Mars imbued multidimensionality in which different aspects of the god appeared depending upon whether the soldier was in the field in a state of war or in the city in the state of watchful peace. Mars was inclusive of both the legionnaire in the field of battle and the sentinel who watches over the peace of the city. Mars was inclusive of the rhythm of the seasons and required purification rites (the *tigillum sororium*) to transmute one from embodying Mars to Quirinus, the deity who represented Mars within the city as watchful guardian (Dumézil *Archaic Roman Religion* 206). The word *martial*, as utilized in this book, seeks to reintroduce and reintegrate the multidimensionality of human existence within and after military service through a useful revivification of mythology.

The simplified fascination of Mars as the god of war has led to the forgetting that Mars has multiple faces, not one. His faces are known as Gradivus and Quirinus. Gradivus is the face of Mars most familiar to us, the god of war. As Quirinus, Mars presides over the life that exists when armor and spear are not donned. Roman lives were a series of endemic conflicts. A person would till soil or sell goods one day, be called to the legion as required, and return to the field life when that particular call of duty was complete. Yet this did not mean the individual Roman was a civilian. There was always going to be another call to the field. This has largely been the human experience ever since the first roots of human civilization occurred. This was a cyclic existence that, as mentioned before, found its natural rhythm in the seasons of nature setting a predictable pattern to a person's life. As such, the god Quirinus comes into their lives as often as Mars. Peaceful Quirinus

> is not the patron, if indeed he ever was, of a social class exempt from military service as opposed to another uniquely warlike group, even in times of peace. [...] Quirinus patronizes merely one of the two modes of behavior which each Roman assumes in turn, [....] "When Mars breaks into rage [*cum saeuit*], he is called Gradivus; when he is tranquil [*cum tranquillus est*], Quirinus. He has two temples in Rome: one, the temple of Quirinus, inside the city, in his character as peaceful guardian [*quasi custodis et tranquilli*]." (Dumézil *Archaic Roman Religion* 261–63)

Mars has two manifestations, Gradivus and Quirinus, two faces to depict the warrior of the field and the protector of the home. "Quirinus is the Mars who presides over peace [*qui praeest pad*] and is worshiped inside the city, for the Mars of war [*belli Mars*] had his temple outside the city" (Dumézil *Archaic Roman Religion* 263). Within the term martial, derived from Mars, there are two faces of watchful peace and energetic war attuned to the seasons of nature. For the

Greeks, Athene, with her connection as defender of the polis, seems to share a great deal of Quirinus energy, or vice versa. The Romans understood there must be a place for Mars to rest within their world, and Quirinus is that deity within which rest and the other aspects of life may exist.

As the Romans adopted and adapted local Etruscan divinities into an amalgam with Greek thought, it is clear that the Romans inherited the concept that conflict contained the energies of the entire pantheon (Dumézil *Archaic Roman Religion* 208–209). Quite often, Roman commanders invoked "a god other than Mars, and sometimes to one far removed from his type" during a battle and surprisingly Dumézil asserts that such invocations were "rarely directed to Mars himself" (*Archaic Roman Religion* 209). This multidimensionality should come as no surprise. Even a cursory reading of the *Iliad* elucidates that Ares does not retain the sole province on the use of nor even inspiration to conduct violence in war. Athene is present and active as would be expected, yet so are a multitude of other Greek deities in play before the gates of Troy. The *Iliad* demonstrates that the Greeks understood that nearly every goddess and god has a seat at the table of conflict. To be a military force and exist in a martial capacity meant to be beholden to nearly every deity, not just the one popularized as "The War God."

In this exploration of Mars, it is clear that to project all aspects of war upon this god's Gradivus aspect, as the monolithic linguistic symbol martial, perverts the understanding of the various modes and diverse energies that are engaged in the life of a servicemember. Mars is not solely a raging berserker, though he can have that quality. Mars embodies courage and sacrifice, acknowledges the human need to purify, and has a place within civil life as a quiet, watchful guardian. Further, Romans saw war as a place for all deities at one time or another. Therefore, the term *martial* in this book encompasses the various aspects of Mars as well as extending to the entire pantheon of deities at play in the multifaceted human existence of a warrior or servicemember's lifeway. It must be remembered that even Vesta, the Roman counterpart of Hestia, who symbolizes the close flame of the hearth, has her place in the small campfire glowing in the night to keep council with a few soldiers staring quietly into the flame as she does burning in the hearth of those soldiers' childhood homes.

In today's discourse, the revivification of Quirinus, which could serve as a beautiful answer from ancestors within martial culture about how to be a warrior when home, has been lost. Our present-day oversimplification has turned Mars into a stereotype rather than a multidimensional complexity of human experience (Dyvik 135). This has significantly imbalanced our views of the martial in today's world. And this lack of balance has sorely affected the way in which modern western societies approach and view those who enter into martial culture through military service. It obscures the fact that once military service is complete, martial culture endures in the civil aspects of life.

While I do use this brief exploration of Greek and Roman thought in connection with martial culture as a way of describing why I believe martial is an appropriate term to describe the multipolarities of the lifeway chosen by warriors and servicemembers, I do not further engage with the mythologies of Greece or Rome in this book for three interconnected reasons. The first is that these two groups of myths are already very well explored in current literature in relation to the lives of servicemembers. The second, linked to the first, is that they have been so thoroughly skewed in terms of emphasizing the elicitation of pathos that any readdress would take considerably more than a chapter to unweave current understandings. The exploration of Mars/Ares along with the associations the Greeks and Romans had with their pantheons and conflict is necessary to strip away the simple concepts moderns have about "war deities," in short, to depathologize Mars. The third and final reason is that Greek and Roman martial experiences, actually representing a small fraction of the total humans across the planet, have confined discussions of the relationship between mythology to servicemembers to a small subset of human population sitting on the Mediterranean. Rather than re-tread the same ground, I am more interested in learning what other martial cultures had to say in myth about living as warriors.

Conclusion

This book seeks to explore selected mythologies through the lens of US martial culture. Martial culture itself is a term that is both introduced and defined here as more effectual and accurate way to approach the lived experiences and social structure of US servicemembers today. A military is an organizational body, while martial describes the soul that, after leaving the body, can become fragmented and incomplete if not understood and recognized for what it is. The servicemember who serves their culture by willingly participating in the discord and confusion of war does not disappear when Mars *Gradivus* calms but must transform to serve Mars *Quirinus*, that of the peaceful protector. The revivification of peaceful Quirinus brings a holistic balance to the martial. Most servicemembers today view the warrior as one who serves beyond themselves. With the brief explanation of why I believe martial is appropriate to describe the holistic societal cultural web servicemembers, veterans, and their families exist in, it is time to move forward.

This book posits a theory of martial culture as a separate but valid lifeway, rich in holistic human experience. But perhaps more importantly, it explores the connections between ancient mythologies regarding martial culture and the contemporary US servicemembers, veterans, and their families. Through the voice of the ancestors, the following pages advocate for a wider awareness of the various mythologies concerning martial cultures and emphasize the need to reconnect with these epical "first stories." The generations that come next within US martial culture should not only understand their place as spiritual inheritors of these myths but have the words, songs, and dances to rebalance thought, feeling, and action thereby encouraging serviceship to act in accord with the ideal values, beliefs, and norms without falling prey to those who would denigrate and violate those values.

Notes

1 *Kosmos*, as used here, is defined by Wilber as including not only the external physical cosmos of the world around an individual but also the "interior dimension of reality" (Mahaffey).

2 Karl Kerenyi explains the androgynous nature of Pallas Athena, "In the case of the name Pallas, one has to associate with a masculine Pallas, with an extraction from androgynous unity; this would explain the androgeny [sic] of the Goddess" (26).

3 The latitude necessarily granted in authority to those in the chain of command include, as necessary, health and wellness inspections of living quarters, limitations on moving off post or base, whether family can accompany personnel to overseas postings, etc. Additionally, the military has its own judicial system under the Uniform Code of Military Justice.

4 Kluckhohn managed to define culture in turn as: (1) "the total way of life of a people"; (2) "the social legacy the individual acquires from his group"; (3) "a way of thinking, feeling, and behaving"; (4) "an abstraction from behavior"; (5) a theory on the part of the anthropologist about the way in which a group of people in fact behave; (6) a "storehouse of pooled learning"; (7) "a set of standardized orientations to recurrent problems"; (8) "learned behavior"; (9) a mechanism for the normative regulation of behavior; (10) "a set of techniques for adjusting to both the external environment and to other men"; (11) "a precipitate of history"; and turning, perhaps in desperation, to similes, as a map, as a sieve, and as a matrix. (Geertz 4–5)

5 The origin of the word derives from the "etymon classical Latin *cīvīlis*" as it relates to matters of the citizen and draws a distinction between the life and status of "the citizen as distinct from the soldier" ("civil"). Civil also characterizes a person as outside of the military sphere as well as serves as a descriptor of behavior "polite, courteous" ("civil"). Civil, as used in the word civilization, intends to apply the life of law, politics, and the behavior of the individual as an ordering principle which opposes the disorderly world beyond the civilization.

6 Possibly the earliest record of this tension is found in the poem of Inanna and Ebih, as recorded by the Sumerian High Priestess Enheduanna (2285–2250 BCE). It describes a mountain who has perfected an ordered paradise in defiance of Inanna herself, she who represents the natural world which includes strife, demonstrated here as a series of destructive natural phenomena (Enheduanna and Meador 91–106). Betty Meador interprets this as a shift from the natural world as good towards a validation of an unnatural other or Eden which is not of this world (109–110). The ordered paradise, an Eden where the lion lies with the lamb can only be described as the ultimate in civility. Civilizations have long worked to dispel natural chaos and deny the cycles of life and death as a constant. Inanna's power is described in terms of elemental forces. It must be remembered that conflict, namely war, was just as much her domain as floods and tempests. The attempt of Ebih to rob Inanna of her power and dominion is the splitting between the peaceful civil and the "Lady of largest heart/Keen-for-battle queen" (Enheduanna and Meador 117). Meanwhile, the Papyrus Lansing, authored during the rule of Pharaoh Senusret III (1878–1839 BCE.), describes the superiority of the scribe and denigrates other trades such as laborers, farmers, and soldiers

(Blackman and Peet 284–98; Shaw 77–78). The scribe, beholden of Thoth, is the learned individual who may pursue a life of letters, undisturbed by the difficulties of martial life. Indeed, this scribe heaps scorn upon those who may voluntarily pursue such a life.

7 The word *martial* derives from the "classical Latin *mārtiālis* which means of or belonging to Mars, the god of war" ("martial").

References

Campbell, Joseph. *The Masks of God: Creative Mythology*. Penguin, 1976.

Campbell, Joseph, and Bill Moyers. *The Power of Myth*. Doubleday, 1988.

Dumézil, Georges. *Archaic Roman Religion: With an Appendix on the Religion of the Etruscans*. Johns Hopkins UP, 1996.

Dyvik, Synne L. "'Valhalla Rising': Gender, Embodiment and Experience in Military Memoirs." *Security Dialogue*, vol. 47, no. 2, 14 Jan. 2016, pp. 133–50, https://doi.org/10.1177/0967010615615730.

English, Allan D. *Understanding Military Culture: A Canadian Perspective*. McGill-Queen's University Press, 2014.

Geertz, Clifford. *The Interpretation of Cultures: Selected Essays*. Basic Books, 1973.

Graeber, David, and David Wengrow. *The Dawn of Everything*. Kindle ed., Farrar, Straus and Giroux, 2021.

Guthrie-Gower, Suzanne, and Gemma Wilson-Menzfeld. "Ex-Military Personnel's Experiences of Loneliness and Social Isolation from Discharge, through Transition, to the Present Day." *PloS one*, vol. 17, no. 6, 6 Jun. 2022, e0269678, https://doi.org/10.1371/journal.pone.0269678

Hannel, Eric. "Veteran Peoplehood: A Theoretical Framework." *Journal of Veterans Studies*, vol. 9, no. 1, 2023, pp. 181–89, https://doi.org/10.21061/jvs.v9i1.397

Hanson, Victor Davis. *The Father of Us All: War and History, Ancient and Modern*. Bloomsbury Press, 2010.

Mahaffey, Patrick. "Evolving God-Images and Postmodernity: Evolving God-Images in the West." Pacifica Graduate Institute, Carpinteria, California. May 2020.

McCaslin, Shannon E., et al. "Military Acculturation and Readjustment to the Civilian Context." *Psychological Trauma: Theory, Research, Practice, and Policy*, vol. 13, no. 6, 2021, pp. 611–20, https://doi.org/10.1037/tra0000999

"Orphic Hymn to the Goddess Athena." Translated by Thomas Taylor, The Encyclopedia of the Goddess Athena, Ray George, 1999, www.goddess-athena.org/Encyclopedia/Rituals/Hymns/Orphic.htm

Shepherd, Steven, et al. "The Challenges of Military Veterans in Their Transition to the Workplace: A Call for Integrating Basic and Applied Psychological Science." *Perspectives on Psychological Science: A Journal of the Association for Psychological Science*, vol. 16, no. 3, 2021, pp. 590–613, https://doi.org/10.1177/1745691620953096

Smith, Michael J. "Can Veterans Experience Acculturative Stress?" *Journal of Veterans Studies*, vol. 9, no. 1, 2023, pp. 103–114, https://doi.org/10.21061/jvs.v9i1.371

Taylor, Charles. *A Secular Age*. The Belknap Press of Harvard University Press, 2018.

2 Martial Culture Theory

Martial culture exists as the lived experience of belonging to a martial *mythos*. Initially, martial culture emerged as a by-product of organizing for conflict. This conflict includes organized violence not only between groups of humans but also between humans and other animals, whether in defending from predators or organized hunting parties.[1] As soon as culture as it is presently understood manifested, martial culture existed.

There has been an evolution to the concept of service in martial culture as human knowledge has progressed over thousands of years. Today, war is not the only defining event for those in US martial culture because many servicemembers have not and will not experience it.[2] What defines martial culture today is the series of rituals, trials, and thresholds that must be crossed. The true challenges are the physiological and psychological changes that are created to bring an individual into a network of highly interdependent tribal groups. The daily life of ritual and symbolism continues to solidify these webs of connection. And the individual measures their personal worth, and is measured by others, based on the level of service they offer their tribal family group.

This familial grouping could be 4, 12, or 44 people; these sizes of human association keeping within the threshold of the ideal number of personal connections a human can maintain (Dunbar 74–84). The servicemember will then move every few years, departing from and merging with new groups; though all will have the common denominators of those within martial culture. Those who fail to live up to the values, beliefs, and norms associated with martial culture in general, and the variations that apply to different units or teams, will suffer from either explicit or implicit separation from the group.

Why Martial Culture?

In nearly every discipline, when the concept of culture is considered in relation to armed forces, it is referred to as military culture. Academic approaches to military culture occur almost wholly through the larger lens of organizational culture, which is normally applied to corporations and other institutions (Meredith

DOI: 10.4324/9781032613222-2

et al.; Voss and Ryseff; Daugherty). The question of whether the role of the individual in the military is a calling, profession, or occupation is largely answered by how "military" is defined (Moskos 23–24). Many civilians and prospective future servicemembers view the military as a government organization that offers careers to individuals. With the emphasis on membership of the soldier in the larger organizational body, the term military aligns with how a secular military organization is described today in terms of a corporate body or government institution. This confusion of terms has informed the approaches of those who study the link between military and culture. Studies in military culture and military sociology, using the approach that the military is an occupation, apply the lens of organizational culture as the most useful model to evaluate culture in the military. According to those disciplines, culture is directly tied to membership within the organization. The etymology of the word military itself denotes organization around a numbering system. One may leave the organization, but the cultural heritage is a lifelong path.

Military refers to an institutional organization to which an individual has a finite membership. That membership begins at time of enlistment and terminates at time of discharge. Military is an adequate term for defining and describing an *organization* consisting of servicemembers, equipment, and structures.[3] Membership in, and influence of the culture, continues to exist beyond military discharge.

The implication of the term military culture is that a person only practices those cultural norms when in the military. Once servicemembers leave, they are no longer in or of the military culture. Thus, military culture implies a finite, temporary, even shallow culture. To quit the military as an occupation is analogous to quitting a large corporation. Cultural norms do not change regardless of whether the person's official status has changed from servicemember to civilian. Military is an organizational term that does not describe the culture of servicemembers because the mythological aspects essential to martial culture are excluded.

Organizations are hierarchies with real power mechanisms that extrinsically dictates an individual's actions through either willing obedience or coercion based on threats of punishment. The term miliary, when coupled with culture, flattens out deeper investigations of human experience and relationality that are intrinsic to the concept of culture. The cultural norms continue to manifest after leaving, leading to difficulties such as communication with civilians or even how to meet basic life necessities (Semczuk; Shepherd et al. 590–613).

Variants of the term military (militarism, militia, militant) often carry negative connotations of a hierarchal organizational structure in areas of civil society where they may serve as a toxin rather than a balm. The continuation of these organizational principles by veterans can be debilitating to the growth of the individual, who, rather than bringing the positive martial cultural values forward and integrating them with new experiences, becomes rigid, inflexible, and stuck in a useless caricature.

I believe military culture is proper to describe these outward organizational cultural studies and this area has provided a great deal of insight into the nature of the services as institutions. In his work *The Square and the Tower,* the historian Niall Ferguson delineates the differences between an outward understanding of hierarchy as the tower of a medieval town, and the unofficial but possibly more powerful network of flatter relationships formed through the natural interaction that occur in the market square of the town center. The theory of martial culture I propose seeks to study and understand the intrinsic human values, beliefs, norms of behavior, and interpersonal connections, which are "flatter" than the official hierarchy and exist and thrive both within and beyond the institution. Martial culture seeks to study the human dimensions of intrinsic value systems and social networks that sustain the servicemember, veteran, and their families through ethnographical, psychological, and physiological approaches during and after the military. Those who leave the military do not leave martial culture, which supports and sustains the servicemember and the military institution.

There is a significant difference between organizational culture and national or societal culture, and the definitions and approaches to each. The societal cultural studies focus on values and behaviors that permeate every facet of life while organizational cultural studies are only concerned with the individual's persona within the sphere of their career. As an example regarding the difference between these two cultural approaches,

> "an individual's efforts to transition between organizational membership— and thus between organizational cultures—is generally a simpler activity than transitioning between societal/national membership. The first simply requires applying for and acquiring a new job, whereas the second typically entails a substantial process of acquiring citizenship. Although in both cases, the change can only occur when both the individual and the acquiring entity agree to the change, the difference here is in the level of time, energy, and effort involved to make the change." (Dickson et al.)

The process of changing one's societal or national membership more closely represents the process to enter and become a full member of the military, as will be discussed in depth in this chapter. Therefore, organizational approaches to the holistic and lived experience of a servicemember/veteran and their families are limiting when trying to understand the totality of that culture. I offer the term "martial culture" as a way of opening up the approaches to studying a single coherent lifeway of servicemembers and veterans through an anthropological ethnographic lens. Rather than an organizational culture, martial culture is a national or societal culture model. I only seek to introduce this theory. And while my own expertise is limited to mythological and ritual considerations, I argue that this overall approach could open a great deal of study in a variety of fields.

From an insider perspective, joining the military is a holistic giving of oneself into martial culture's power. This deeper individual connection and identification with cultural norms indicates that martial cultural values, beliefs, and norms will continue even after leaving the organization. In short, one can leave the military, but not the culture. Parallel work has been conducted on this phenomenon of acculturation in several very recent studies which are beginning to recognize the cultural nature of service and the acculturation processes that occur upon entering martial culture (Shepherd et al. 590–613; McCaslin et al. 611–20; Olenick et al. 635–39).

Loss of martial cultural identity for veterans is a mistaken concept. A potential analogy is how one may be born and raised in the organizational nation-state known as Guatemala, yet once they emigrate to another country that person is still holistically culturally Guatemalan. The term *martial culture* encapsulates the values, beliefs, and norms of behavior attained through serving within a military organization to which the participant continues to adhere to or be affected by, consciously or unconsciously, throughout the remainder of their lives. This experience can vary in degrees, just as an immigrant can work to throw off the old cultural identity—which is different from national citizenship—and adopt the values, beliefs, and norms of a new nation (Okafor and Kalu). Emigrants are still, in a very fundamental way, beholden to their original cultural heritage. It is a part of them, and they are a part of it. Cultural influence extends beyond the termination of membership to the organization or institution.

Martial Culture Defined

Martial culture exhibits differentiation in the use of language. Words are symbolic "representations contain[ing] information about sensations, perceptions, concepts, and categories" (Siegel 227) with limits, either manifested in sound (voice), movement (sign or body language), or as witnessed here, scratched patterns (pictographic or lettering). Words also "move beyond the physical world and link the mental representational worlds of separate people [....] Human language permits information processing to be shared across individuals" (Siegel 227). Unfortunately, the receivers are subject to their own knowledge and ignorance as they extrapolate meaning from words. Therefore, a text heard/read by two different individuals can have radically different interpretations due to the variances in the individuals. How a word-as-symbol is defined or interpreted is related to the worldview and life experience of the individual who receives it, which may not be the intended interpretation of the individual who used it. In other words, language can be a barrier for two cultures who seek to find common understanding (especially if they both use the same words for different purposes); therefore, clarification of terms is necessary. Misconstruing the meaning of a word, especially between cultures, can lead to confusion, misunderstanding, and hostility, especially if an assumption exists that both parties are supposedly from the same cultural cloth.

In the United States today, the martial culture that supports the military exists as a secular, ancestrally oriented, nomadic, totemic, chiefdoms of tribal social groups. This martial culture maintains historic and contemporary personalities as guides to behavior and is heavily imbued with symbolism. The individual exists to support the whole and gains self-worth from the same; with loss of membership to the tribe resulting in the loss of self-worth. US martial culture is hierarchal with internal caste-like rules of association where every member of the society is a potential regulator of behavior. While often overlooked, these cultures include the family of the servicemember who also exist within martial culture. The lifeway is often nomadic, as members often go through mandatory relocation every three to four years punctuated by deployments, missions, training, and education which contributes to a constant state of liminal existence. And finally, it is extremely totemic, as symbols or objects are infused with a type of sacredness. These may include unit symbols, personal symbols, or objects adopted as sacred by the individual through their experiences in martial culture, and in many cases the US flag, for many reasons that are extremely personal and visceral (Rodriguez, France, Osuna, Blackmarr, McCoy, Skiles).

The US military organization has been secularized in order to conform to the government's requirements. Because the military is a subordinate institution under the US government, which practices separation of church and state, it is required to emphasize and demonstrate "professional government employee" image. However, time spent within the military exposes the underlying human-to-human and human-to-myth connections that sustain and allow the military to function as an organization. The extremes of experience that servicemembers and their families face demands a unique mytho-cultural fabric. While official government records usually do not capture these numinous experiences, many servicemembers and veterans share stories that routinely describe mythic states of existence. Military service is not a secular government experience but a deeply human one, replete with multidimensional mythological power. While many mythologies and rituals are still practiced, they are hidden and stripped of their power when referred to as training, traditions, and customs. Because of this, myth has gone underground in the military. Emphasizing logos over mythos has resulted in an unbalanced culture. The technique of hiding myths through turn-of-phrase makes martial culture palatable to a public clamoring for secularism in their government offices. Yet, it does nothing for servicemembers and their families who live the myths in secret and then wonder why, when they choose or are forced to leave the military institution, they feel and act so differently than the civilian cultures they have been thrust into.

Today the mythological dimensions of martial culture are misunderstood due to a combination of secular lenses, prejudices, and faulty assumptions accumulated over millennia. The evolution and appropriation of certain terms, which have led to stereotyping servicemembers, contribute to this misunderstanding (Okafor and Kalu). The largest contributor to these misunderstandings is that civilian cultures have evolved and grown to become separate and larger entities. Throughout the world, and particularly within the United States, civilians make up the overwhelming majority of people. As such, cultural groups have established a standard baseline of values, beliefs, and norms of behavior that are defined as "normal" within a given spectrum. This norm is actively described in moral and ethical terms which explicitly defines deviations as morally negative. Martial cultures, as differentiated independent cultural groups with their own codified values and behavior, are viewed as a deviation from the spectrum of normal as defined by the dominant civilian cultures.

An exploration of mythology that speaks to the martial culture must account for a wide variety of preconceptions. The mythologies today are largely viewed as metaphors or fictional stories and seen as the sole property of religious practitioners or those who specialize in the liberal arts, social sciences, and psychology. There they have flourished and found life through application to metaphorical conflicts that appear in the psyche and soul of individuals. The preservation of myths by the religious and academic communities is a godsend in that they have kept the myths alive and have found use for them. Unfortunately, the myths have largely been forgotten by the militaries of today. While myth and ritual are present in the military, they are faded relics of their former selves and not recognized for what they are. The rituals have been reduced to customs or traditions whose true meaning has long since been forgotten.

The Greeks, Romans, *samurai*, Norse, Hindus, and the *Diné* (Navajo) are martial cultures who maintained active dialogue with their rituals and mythologies. Myths served several vital functions to buoy and guide the individual and collective psyche and soul of those in martial culture. Given the numerous psychological, ethical, and moral conflicts suffered by the active servicemember and veteran community today, the time to rediscover the power of these mythologies is now. The civilian culture's emphasis on tragedy and despair dominates today's view of servicemembers, veterans, and their families. Stripping away many of the aspects of a beautiful lifeway of service, courage, and familial bonds created in the face of extreme human experience. Further, if civilian culture acknowledges the validity of martial culture, the myths can serve as a way to help connect civilian cultures with martial culture. But perhaps more importantly, the stories can reintroduce the existing servicemember to the very real heritage they share and, in so doing, inculcate a greater wisdom, a balanced ethos, and a genuine respect and recognition of their connection to something beyond themselves.

The roots of martial culture lay in the collective human experiences of serving through use of force. Those experiences coalesce into a series of values, beliefs, and norms of behavior. Mythologies emerge out of accumulated collective lived experiences as a way of supporting the culture and the individual. Mythology is inextricably bound within culture, directly shaping the formulas a culture uses to define itself. Martial culture began at the intersection of the necessity of serving through sacrifice and the development of sense-making individuals need to explain their experiential interactions with the world.

Both military organizations and the martial cultures that sustain them are based on previous evolutions of each. A military organization is an amalgamation of preexisting organizations and the same is true of martial cultures. Martial cultures, like other cultures, have split, reformed, and gone through countless iterations of alchemical processes to refine themselves into variations of culture that seem appropriate to the people, time, and place.

Intertwined Roots

Martial cultures, though influenced through a variety of local or regional customs, tend to have commonalities. Martial cultures are modeled on and inheritors of the various cultural influences that came before. The US military and underlying martial culture began as an amalgam of British, French, Prussian, and Native American military practices and martial cultures. Many militaries share underlying martial cultural values: "Duty, discipline, and selfless service are traits often shared by combatants on both sides of the battlefield" (Mansoor and Murray 18). As has been repeatedly noted by scholars and historians, martial cultures tend to share similar bedrock values.

The *Handbook of the Sociology of the Military, 2nd edition,* approaches study of the military from a sociological method, yet still uses terms such as organization and institution, with more emphasis on organizational history. Yet this field does come closer to identifying the realities of military culture as distinctly separate from civilian culture,

> [E]ven among the special sociologies, the one dedicated to the military seems to be "especially special". For centuries, the military world and the military mind-set have constituted a quite different, quite separate environment from the other institutions, groups and aggregates of civil society, and in part they still do." (Caforio et al. 4)

Further, it is noted that the military is so different as "to require, on the one hand, an adequate sociological preparation [...] and on the other, thorough, possibly first-hand, knowledge of the particular study environment, that of military society" (Caforio et al. 4). The assertion is that to really be a military sociologist, one must have served in the military. This argument asserts just how removed martial cultures are from civilian cultures. Much of the academic discourse today within sociological circles is on the fractured relationship between the civil and the martial, referred to as *civil-military relations*. The issues are many, but at the heart are the fundamental differences between the *civil* and the *military* as previously discussed.

The cadre that usually form the core around which a new military organization emerges carry with them the cultural trappings of the martial cultures they previously belonged to. In moving back in time to explore the martial cultures of those who taught and aided the burgeoning new military organizations, a web of cultural and organizational influences emerges that stretch back to prehistory.

A fully formed martial culture is consciously aware of the warrior who provides for and protects the social order by valuing their life as subservient to the larger society. Because warrior is so ubiquitous in language and cannot be easily separated from contemporary martial culture or current translations of myths, in this book the word warrior refers to a member of martial culture who protects and provides for the collective society—risking the welfare and existence of the individual in service to something outside oneself—and is inclusive of all genders. In this way, warrior and servicemember are interchangeable terms with the notable distinction that in many of the myths, warriors fought as individuals whereas today a US servicemember/warrior exists and operates as part of an interdependent team. This definition of warrior engenders a more holistic understanding, and as such is inclusive of servicemembers who may not see combat (Parker et al. 12). Yet the trials and deprivations that are endured to gain admittance to, and live within, US martial culture are largely universal. All servicemembers fulfill the "provide for and protect" function of the warrior as described herein. That service takes many forms, though it can include the loss of one's own life as well as causing the death of another. A key factor for a functioning martial culture is the conscious awareness and acceptance of the life-long commitment to serve as a warrior. Native American, Norse, Hindu, and Japanese martial cultures were just a few that, at one time or another in history, inculcated a recognition that martial culture is a lifeway and not a brief term of service to be switched on and off with secular industrial precision. A largely unacknowledged part of that service is the acceptance of the life-altering changes that occur as a result of serving in the military. Campbell saw this transformation as a complete rebirth, in *The Power of Myth* he observes, "You've undergone a death and resurrection, you put on a uniform, you're another creature" (1). Being wholly transformed into another creature causes influences far beyond a term of service and accepting the changes is an example of what martial culture demands as a life-way.

The term *veteran*, used to describe a subset of civilians who are no longer in the military, is unique. Many nation-states today who practice compulsory military service have no need for a word that differentiates a civilian from a civilian who once served in the military. William Meadows illustrates that the Kiowa social order was one that existed to guide a person from child to old age, and martial service was inseparable from the entire life of an individual (Meadows 3–7). Likewise, the Chiricahua Apache Chief known as Cochise never ceased being a warrior in service to the Chiricahua, and his people would never have asked him to. The US model of veteran implies that there is a recognition that martial culture exists beyond belonging to the military organization.

Many veterans today feel that their association with martial culture must be suppressed and hidden. They find themselves instructed by influencers such as Transition Assistance personnel, family, and wider society to act like a civilian. The result is cognitive dissonance that is damaging to the individual. A true understanding of martial culture allows for unashamedly integrating the service-member aspect of the individual into life beyond formal membership to a service branch.

The focus of this book is the mythical dimension that defines and supports martial culture; military organization is only discussed insofar as it impacts that culture. For example, a military organization that uses conscription or has very brief initiation rituals will have a markedly different culture than one that only accepts volunteers and has extended initiation rites. Therefore, an examination of how the US martial culture evolved is key to understanding why it exists in its current forms today. As Boas notes, a person cannot "attain full significance [of the culture] without knowledge of the historical development underlying the present patterns" (qtd in Lévi-Strauss 9). Exploration of the historical development of martial peoples demonstrates the need to recognize martial culture, which in today's secular world is a shattered specter of what it needs to be. To fully embrace this reality will give purpose to veterans and will also remind them of the life-long responsibilities to live the ideals of the selfless warrior. Lastly, acknowledging the existence of martial culture will begin to heal the rift that exists between the martial world and the civilian world.

The mythologies of the migratory era of Homo sapiens demonstrate the earliest roots of martial culture. The use of force to change the environment into something more suitable for foraging marks the beginnings of martial culture. In the earliest stages, nomadic Homo sapiens groups were too small in number to survive unless every member was involved in utilizing force to defend, attack, procure food, and/or shape the landscape. Each member was constantly at risk of losing her/his life to other animals and people while also required to either inflict force or enable the infliction of force on other animals or peoples. All Homo sapiens were members of, or belonged to, a martial culture out of collective necessity.

The Beginnings of Civilization

As the first cities, supported by agriculture and defended by fortifications, began to emerge, a new type of Homo sapiens experience came into being. The advent of specialization allowed for some people to avoid the experiencing or utilizing of force (or risking danger to their person) in support of the community. As communities grew larger, those with unique skills or trades became such exceptions. In many cases, these exceptions to martial culture made valuable contributions to the overall society. While the majority of the population of these early civilizations make up the earliest armies, the ruling class was likely the first to experience this unheard of and all-together new experience and role: the civilian.

Civilians would remain an extremely small number compared to the overall size of the population. From the falls of Jericho and Troy to the warring states of China and the civilizations of the (pre-European contact) Mayans, history and archeology demonstrate that a constant state of endemic conflict existed throughout human existence (Schmookler; LeBlanc and Register). In mobilization for war, a majority of society would be pressed into military service.

Certain societies are famous (or infamous) for their warrior castes: the Greek Spartans, the Aztec Jaguar and Eagle warriors, the Japanese Samurai, and the knightly class of medieval Europe are notable examples. There is an assumption that those who belonged to the warrior castes were the only ones inducted into conflict. The reality is the highly skilled warriors served as a core cadre or vanguard; the common peoples in the society were pressed into service to make up the rest of the army around the core group. Their families would be the forerunners of today's military families.

Therefore, to be alive and (usually) male was to be drafted into a reserve force and called to military service as deemed necessary. The only exception was if one had enough affluence or was too impaired to march. The bulk of the armies and navies were populated with part-time soldiers and sailors who did not have a choice in their participation. Yet it must be remembered that those in the warrior castes were not necessarily volunteers either. Caste was usually hereditary and those who were born into warrior castes had little more agency than conscripts themselves.

Gender

Just as veterans are excluded from military organizations, female-identified servicemembers are largely omitted from history. The resultant binary has also translated into a widely held belief that the masculine archetype traits form the seat of all aggression. While there is abundant evidence that men have been the primary participants in martial culture, existent but suppressed evidence indicates that women had a much more prominent role in military organizations and martial cultures of the past.

Possibly the earliest divisions of labor were by gender and by age. However, that division did little to restrict any gender from martial culture. Whether a threat was from marauding animals or other groups of humans, necessity superseded exclusively gendered roles involving the use of force in service to the wider community.

A number of theories regarding the evolution of labor division reframe the role of women in the social function of martial culture as a full member of the group. "Most researchers currently reject the idea that a gendered division of labor emerged in the Pliocene, at least not in its modern form [...] the typical patterns of labor division emerged relatively late in human evolutionary history" (Kuhn and Stiner 953–54). While the study does not extend this concept to defense of the community, it is likely that no such gender barriers existed in Middle Paleolithic communities as modern society understands them today. If so, then the common extension of activity that connects hunting to conflict in other societies would reasonably exist in the Middle Paleolithic. Women as equal participants in hunting, along with early myths such as the poems of Inanna that portray what our inherited twentieth-century psychology calls the feminine as a war goddess, one can begin to reimagine the role of women in the earliest formations of culture.

There are several issues that contribute to the current understanding of gender and conflict. One is the development of social norms that contributed to encouraging conformity of exhibited traits such as passivity in women and aggressiveness in men. What may have begun as a potentially well-intentioned symbiotic relationship (gendered division of labor) became a codified, socially constructed "normalization" process. It is possible that socially enforced norms for behavioral traits in men and women informed sexual selection, thereby attempting to pass on certain socionormative tendencies to future generations (Little et al. 366–75; Street et al.; Lyon and Montgomerie). Holistic redefining of gender roles in relation to service in martial culture is necessary, not because women have never served as warriors, but because women are largely underrepresented in service. A potential contributing factor to this underrepresentation may be the result of a person's earliest socialization of what is considered "proper" for each gender in regard to serving as warriors (Martin and Ruble 353–81; Brown et al. 202–19).

The social attitudes of gender roles in conflict have not been uniform. In many societies, there has been a concerted effort to downplay the roles women played as active combatants. In fact, the success of that suppression has been so far reaching that it has been largely adopted by some of those who champion women's rights. Some advocates describe females as the peaceful gender and thus any woman who does participate in conflict as a combatant has taken on a "man's role" and is confused. Pamela Toler asserts that women are not natural pacifists by presenting three examples of women warriors: "Tomyris, the warrior queen of what is now Kazakhstan" who defeated Cyrus the Great (530 BCE), the Iceni Britain Boudica who led a revolt against the Romans (49–50 CE), and "Lakshmi Bai (1828–58), the Rani of Jhansi" who as a widowed ruler led her kingdom to war against the British (19–31). These three examples are unique in that they have been allowed to exist in various records. For the few known examples, there are probably many more women who have been left out of recorded history.

Biological sexual selection (mating preferences) may have contributed in some small measure to differentiations between male and female. I assert it is the suppression of the woman-as-warrior that has been far more damaging to an understanding of Homo sapiens and conflict. This suppression affected, and is still affected by, the ordering of the attributes assigned to masculine and feminine when Jung defined archetypal energies, much of which was informed by his socially normative Western understanding of male and female behaviors and roles (Bjorkman; Martin and Ruble 353–81; Brown et al. 202–19).

Archeologist Jeannin Davis-Kimball notes that some of the ways in which people derided Herodotus's accounts as fantasy rather than historical descriptions were to attack his seemingly outlandish tales of events, in particular, those of women warriors (Davis-Kimball and Behan 52–54). "Only in recent times have archeologists' findings begun to corroborate many of his seemly outlandish contentions—including those regarding women warriors" (Davis-Kimball and Behan 52). For example, "roughly 20 percent of the Sarmatian warrior graves excavated in the lower Volga region belonged to women, with bows and arrows being the most prevalent weapons" (Davis-Kimball and Behan 54). Further, an archeological find, dubbed the Issyk Gold Man was later revealed to more likely be a woman (104–107). In conversation, the anthropologist who examined the remains admitted that the bones "were very small and could have belonged to a female" but that the presence of "prestigious artifacts, particularly the sword and dagger" probably swayed the discovering archeologist believe it was a male (106).

The number of historically accurate and now increasingly verifiable accounts of women in battle is staggering. Especially as many women, like Mulan of the Chinese legend, pretended to be men in order to join the martial culture. In US martial culture women lived and died largely forgotten, such as "Deborah Sampson, who served for 17 months in the Continental Army during the Revolutionary War as Robert Shurtliff, and Lucy Brewer, who served with the Marines aboard Old Ironsides as George Baker during the War of 1812" (Schulte). To many people's surprise, "historians have found that an estimated 400 to 1,000 women, perhaps more, disguised themselves as men and took up arms in the Civil War" (Schulte). Those numbers may seem out of proportion, however, with the covering up of evidence by both men who viewed martial culture as a "male-only" world as well as women who were trying to hide their involvement.

Statistical extrapolation hints that the number of women known to have participated in conflict is a small fraction of the total number of women warriors who have existed in martial culture. As David Jones asserts, the

> historical record shows no martial domain exclusive to either males or females [....] For every woman warrior that is known, thousands, perhaps millions, escaped detection [....] How many women who disguised themselves as men to enter war succeeded in their charade? The accounts of those women whose sexual identity was finally revealed showed them to be unusually courageous and intelligent, in some cases able to fool their male counterparts for many years. It seems evident that many more such women remained undiscovered. (249)

When social conditioning and culture began to disallow women as warriors, some began hiding their gender in order to participate as combatants. When militaries began to institute actual physical screenings (the US military began them in 1872), the ability of women to serve in secret was severely reduced (Schulte).

Two *Diné* women may have been the first to openly enlist as women in the US Army. Amateur historian C'de Baca, while researching records from the American Indian wars, discovered two Navajo women,

> Mexicana Chiquito (whose given name was Nal-Kai) and Muchacha—who were enlisted as Army Scouts by the 20th Regiment, US Infantry, at Fort Wingate [...] Mexicana Chiquito, 24, served from May 24, 1886, to Oct. 11, 1886. Muchacha, 21, served from May 26, 1886, to Oct. 11, 1886. (Brunt)

This discovery illustrates how far back Navajo women served in combat roles as scouts in the US martial culture and even uncovered that "Mexicana Chiquito had applied for and received an Army pension" (Brunt).

Today the gender barrier in martial culture is finally dissolving. Today, fathers may stay home while mothers deploy to combat zones. David Hay asserts, "The assumption that war is something essentially male—be it the apotheosis of masculinity or the incarnation of patriarchy—has banned the study of the female combatant to academic purgatory" (Toler 5). Toler describes this problem as a subset of a larger push in history, science, and other fields to minimize or erase records of female as warrior (5). The belief that women are naturally disconnected from conflict and combat has yielded interesting allegiances between feminists and conservatives. Activist and poet Grace Paley asserts that "war is man-made. It's made by men. It's their thing, it's their world, and they're terribly injured by it," while military historian and theorist Martin van Creveld argues that "women have played in war, namely as its causes, its objects and its victims" (Toler 8–9). Hence, Toler asserts "you will find no unbiased observers. Advocates on both sides are guilty of special pleading, cherry-picking the evidence, and presenting opinion as irrefutable fact" (8). The arguments from feminists, on the one hand, and conservatives, on the other hand, seem intent on ignoring the need for a detailed and in-depth study of women as combatants in favor of perpetuating socially reinforced norms. The efforts to stereotypically feminize women and masculinize men have resulted in a wholistic misunderstanding when it comes to gender roles in general, much less how those roles engage with war and conflict.

To fail to recognize the commonality between genders as equally capable in the application of force is to heap prejudice on women who have served in martial cultures as warriors. Leisa Meyer observes that failure led to demonization of women who joined the Women's Army Corps in World War II. While women were already suffering from Army attempts to isolate their roles to "women's work" positions such as kitchen/cook staff and laundry units, these servicemembers also suffered slander and were maligned by both men and women as cross-dressers, sexual deviants, prostitutes for male officers, and women who had abandoned their families (Meyer 81, 33–50; Treadwell 47–49; Bjorkman).

Women as warriors should come as no surprise to those who study mythology. Warrior goddesses are as ancient as warrior gods. Mythologies of martial cultures demonstrate gender is not a determinate for belonging to martial culture and military organization as Toler observations demonstrate. However, class and emergent social norms have long worked to make it one. Because of this unremembered heritage, humans trended away from recognizing women as active combatants as civilizations came into being and grew.

Collective versus Individual

Although American society promotes romantic individualism, which can accommodate a voluntary self-identification with a group, its fundamental belief that the individual does not submit to the culture is in direct contradiction to a core value of its military. The principle upon which the martial and the civilian cultures are split is: which takes precedence, the collective or the individual? Martial culture is a collective in which self-sacrifice is the cornerstone of the culture. As Sapolsky notes,

> Implicit in the very nature of the contrast are markedly different approaches to the morality of ends and means. By definition, collectivist cultures are more comfortable than individualistic ones with people being used as a means to a utilitarian end. Moreover, moral imperatives in collectivist cultures tend to be about social roles and duties to the group, whereas those in individualistic cultures are typically about individual rights. (Sapolsky *Behave* 501)

The typical rights of an individual are stripped away upon entering martial culture. The martial culture demands the individual embrace their role and the duties required by the collective.

Sapolsky also highlights studies that conclude that collectivist societies use shame (external motivation) as the method of correcting behavior whereas guilt (internalized motivation) is emphasized in individualistic cultures. "Shame requires an audience, is about honor" Sapolsky writes. "Guilt is for cultures that treasure privacy and is about conscience" (Sapolsky *Behave* 502).

Today's US martial culture is a hybrid of the two. Shame is used with the intention to eventually be internalized as guilt so that, even if no one else is present, the member of the collective will act in accord with the social order. The collective aspect of martial service does not dissipate with the move from shame-motivated to guilt-motivated actions in today's martial culture. The internalization of the laws, both formal and informal, is designed to provide a guide to behavior.

The individualistic society in the United States today is highly ambivalent towards both collectives and its military. The mistrust of submission to a collective culture is inherent in the founding principles of the United States and stems from the mistrust of a standing army whose excesses, endorsed by the British Crown, were causes for revolution. This resulted in some of the earliest decisions on the formation of military organizations. Throughout its history, the United States has been very utilitarian in its approach to militaries, with a great deal of angst regarding a standing army versus raising one as needed. In the beginning of the US military, armies were only formed for fighting wars, but standing navies were considered a constant necessity. Martial culture was never in danger of being abolished in the United States.

Perspectives on Martial Cultures

In the United States' all-volunteer military of today, which is quite different from the mixed volunteer/conscript militaries that composed the majority of American's past, the old saying that the US military is a slice of America is no longer accurate. The evolution of martial culture in the United States, after the dissolution of the draft in 1973, has accelerated away from civil culture, and vice versa. For the US martial culture and military organization, the evolving social systems have perpetuated differences between servicemembers and civilians in many ways.

Seventy-five percent of Americans ages 17 to 24 do not qualify for the military because they are physically unfit, have a criminal record, or do not meet the minimum education requirements (Theokas 1; "Ready, Willing, and Unable to Serve" 1). A 2015 report indicated that, on average, servicemembers held a higher education level than their civilian counterparts (Department of Defense 39–40). The heightened requirements and education levels are partially the result of the military inculcating a culture of continuous advancement and self-improvement. Servicemembers are evaluated not only on proficiency in their craft but also in other areas, such as education and community involvement; that scrutiny has facilitated a type of servicemember who is vastly different than the stereotype many believe (Schmidt 13–24).

There is a more appropriate model to describe the dilemma of veterans and their families; that of the immigrant (Coll et al. 488; Waters and Pineau 2–3), for it is not only the veteran who finds themselves adrift. Their spouses and children also find themselves within a different culture, seeking some purchase to avoid an emotional free-fall. This dilemma is compounded for spouses in particular, who, having never gone through the intense rituals that the servicemember experiences, is in many cases not fully integrated into modern martial cultures. Therefore, expectations that "going back" to a civilian culture will be easy belie the reality that the spouse occupies a liminal space between worlds yet is still very much changed from their connection as a family member.

Today the American military and the civilian population they serve face each other across a chasm. That chasm does not simply disappear when the warrior and their immediate family are flung into civil culture (Cox et al. 1–2). Service-members and their families move from the military organization into American civil culture. The values and behaviors commonly associated with Americans are summarized in this list compiled for visiting foreign exchange students. The first, asserts that, "Americans strongly believe in the concept of individualism. They consider themselves to be separate individuals who are in control of their own lives, rather than members of a close-knit, interdependent family, religious group, tribe, nation, or other group" ("American Culture: Study in the USA"). In contrast, US martial culture emphasizes the good of the team over the individual. The article goes on to teach that, "Americans believe that all people are of equal standing, and are therefore uncomfortable with overt displays of respect such as being bowed to" which contrasts with the hierarchy, both formal (rank) and informal (advanced schooling/experiences) that delineate servicemembers position and value within the culture ("American Culture: Study in the USA"). And finally, the article insists that the

> belief in equality causes Americans to be rather informal in their behavior towards other people [....] Many people visiting the US are surprised by the informality of American speech, dress, and posture. Don't mistake this for rudeness or irreverence; it's just a part of their culture! ("American Culture: Study in the USA")

The insistence in formal address, uniforms, and proper posture inherent in US martial culture again demonstrates the cultural differences between civil and martial culture. Descriptions of typical American culture, like the one described above, discuss a cultural value system different than that of the US martial culture, where the individual derives meaning from the group, salutes and uses specific ways of address, enforces hierarchical roles, and insists on specified dress codes that are used to evaluate the individual as they relate to their role in the culture.

The popularity of organizations such as Team RWB, Team Rubicon, and many others demonstrates how, like immigrants, many veterans voluntarily cluster socially ("Mission."; "Built To Serve."). Additionally, various organiza-tions that use both programs and rituals to "treat" veterans and their families are believed to help reintegrate veterans into civilian society (Tick; "Home"). These programs reconnect veterans to other veterans, reestablishing the social collec-tive and thereby healing the wounds of excommunication by providing support and direction.

Veteranism describes a phenomenon that was stumbled upon during a medi-cal study (Markel et al. 2). The study seemed to illustrate that veterans identify as veterans before identifying with other personal characteristics such as race

(Walsh et al.). The rise in veteran-owned businesses ("Facts on Veterans and Entrepreneurship"), as well as certain cities or communities that are disproportionately populated by veterans and their families, also tends to demonstrate this immigrant phenomenon ("Cities with the Most Veterans."; Waters and Pineau 5–6, 208). Further, there are suggestions that veterans may give preferential treatment to each other over civilians if they know nothing else about the individuals they are interacting with. This is demonstrated in psychological treatment, where servicemembers overwhelmingly prefer a psychotherapist who is themselves a veteran, if for no other reason than the language used in describing their experiences is so different that someone unaffiliated with the military will simply not even be able to participate in the dialogue from the outset (Johnson et al. 1–2).

Economic advantages, and perhaps increased comfort with other veterans and active military personnel, appear to be driving those in martial culture to cluster socially and economically in an effort to reconnect culturally ("Best for Vets Places to Live"; "Cities with the Most Veterans."). The trend of martial culture becoming more remote from civilian culture, which is largely unaware of the former, continues.

Many veterans find that their past professional skills are inadequate in the civilian cultural setting and therefore must gain civilian credentialling to allow them to enter the civilian workforce, regardless of their actual experience, knowledge, and skills from military service.

Similarly, immigrants entering a foreign country's professional settings face similar issues such as "Professional identity of immigrants is discounted, minimalized or stripped away" and workplace "discrimination and non-recognition of foreign education and experience reduce career satisfaction" (Okafor and Kalu).

US martial culture—like any other culture—is filled with interpersonal conflict. While many veteran organizations would prefer a show of solidarity, the reality is that like any band, tribe, or chiefdom, there are real human interactions, disagreements, and outright disharmony between individuals and even groups. To believe that belonging to martial culture is to somehow participate in a blissful oneness with all like-minded people is to ignore the reality that each society has its internal divisions populated with people ranging from saints to criminals.

The Current US Military Organization and Martial Culture

The US military today is not a singular monolithic entity, but rather a collection of chiefdoms composed of several tribes and bands. Each chiefdom, tribe, and band has individual cultural nuances that vary widely within the larger umbrella of martial culture. To give an example of just how many different variations of the US martial culture there are, the following describes just how many different military organizations there are.

All members of the US armed forces enter martial culture the same way. And in one of the most unique situations throughout human existence, they all voluntarily do so. Each service branch is similar to a chiefdom. Each branch has a series of installations (bases, posts, stations) which are analogous to dispersed villages. Some are quite small, but on the average servicemembers and their families can live, shop, and work within most of these installations. The presence of families is largely the norm, except in combat zones or remote locations. As a Pew study summarizes,

> Today's military is smaller, older, more diverse and more likely to be married than the force that served a generation or two ago. A larger proportion of minorities and women serve as officers and enlisted personnel [and] more of America's fighting forces are husbands or wives—and a growing proportion is married to someone else who serves in the military. ("War and Sacrifice in the Post-9/11 Era" 73)

Military installations have their own schools and daycare facilities for children, as well as hospitals, playgrounds, and even golf courses. However, many families choose to live off of the base or post when possible. Living off base is a popular option, even when stationed overseas in a foreign country. And while this choice allows for outreach my martial culture into civilian communities, in truth the day-to-day life of a servicemember is very much intertwined with martial culture. The installations they live on or near are small towns, with their own law enforcement, fire departments, and are even governed by a separate legal code that applies to all servicemembers and in many cases, their families when on the installation. The installations provide jobs to many of the civilians in the area, giving rise to the notion of "military towns." Often coined as the "military family" by servicemembers, these installations become the epicenter of an isolated and diverse population within the larger society "outside the gates."

One of the diverse elements that has come about in recent years is also the number of same-sex marriages in American martial culture. For these couples, the installation may be the only safe haven to be freely open as a family since many installations may be in states or countries that do not recognize same-sex marriages. These villages are in many ways home, and many families find they miss certain locations so much they end up moving there permanently after they leave the military organization.

At the same time, especially with the constant rotations overseas, servicemembers and their families are consistently exposed to new cultures and peoples. At a typical military installation, regardless of location, one will hear multiple languages and have access to a diversity of foods for several reasons. One is that people do not need to be citizens of the United States to join the military. There is a rich history of citizens of other countries who volunteer to serve (Batalova and Zong; "Immigrants in the Military: 5 Things To Know"). Another is that when servicemembers are stationed overseas, many marry foreign partners who bring with them another layer of martial cultural diversity. This results in base grocery stores (commissaries) carrying far more diverse food selections from around the world. US military installations are cities and towns made up of holistic and intertwined communities.

The communal feel and preference for finding others who are of martial culture, especially with regard to the rising trend of couples in which both spouses are servicemembers or one is a former servicemember, has given rise to an increase in multigenerational service. Tribalism has a great deal to do with family relationships, kinship selection, and sexual or mate selection. In this way, preference for those connected to martial culture manifests in actual blood ties. Amy Schafer notes that more

> than 25 percent of new enlisted recruits have a parent who has served in the military. When the aperture is widened to include a connection to broader family members, such as an aunt/uncle, cousin, sibling, or grandparent, over 75 percent of new recruits for each service have a family member who has served [.... This suggests] that recruits with a direct military family connection are significantly overrepresented in today's U.S. military [....] 45 percent of active duty, 47 percent of military spouses, and 57 percent of veteran respondents have a parent who served. Even more strikingly, 53 percent of their survey respondents had two or more immediate family members who served in the military. Indeed, the military family connection may be more significant than any other variable in determining propensity to serve. (Schafer 6)

Surveys have demonstrated that even when no blood ties are apparent, the concept of "military as family" is a nearly universal constant (Ahern et al. 4–5) and other studies have explored a potentially significant correlation with genetic influences on enlistment and lifetime service (Beaver et al. 1). And as noted earlier, according to the Department of Defense "11.8 percent of Active-Duty marriages are dual-military marriages," meaning that servicemembers are increasingly looking inward for not only kinship selection (friends) but also sexual selection (spouses). As Amy Schafer phrases it, "If interaction with veterans is one of the best ways to encourage future service, then it is unsurprising that military service is perpetuated within military families" (Schafer 11). The number of factors that will increase the tribal-ness in a very anthropological sense, to include blood connections, only increases the remoteness and separation of martial culture from civilian culture.

In terms of collective values and communal regulation, the military of today is still "stuck in amber," having more in common with older sapiens social models composed of bands and tribes (Dunbar; Hanson "Victor Davis Hanson – War and History, Ancient and Modern"). Many militaries throughout the ages have found it necessary to recruit a more diverse population than those they served. This was increasingly the case as civilizations grew in wealth and citizens found it more convenient to hire their protectors rather than serve as their own. The American martial culture is no exception. Investigative journalist and author Thomas Ricks asserts that today's military exists at the intersection of conservative account-ability and liberal diversity.

> Violation of either code will damage a soldier's career. Undergirding this two-part approach is the sturdiest social safety net in America. The combina-tion works [....] The Army is [....] almost a Japanese version of America— relatively harmonious, extremely hierarchical, and nearly always placing the group above the individual. (Ricks)

Martial culture is different than civilian culture and is organized around and sup-ported by rituals and myth primarily tied to sacrifice and service.

According to mythologist Joseph Campbell, rituals heavily reinforce the ped-agogical and sociological aspects of the culture (Campbell et al. *The Power of Myth* 31). Within martial rituals, there are certain mystical aspects associated with particular ceremonies. The decision to engage or suppress them in a martial experience is left to the individual participant. The individual attempts to deter-mine how the various rituals experienced within martial culture will influence them in that arena on a conscious level. On an unconscious level, they work to affect similar changes one sees in those who have experienced the *participation mystique et fascinans* in the various rituals.

Ritual

Mythology consists of the cosmological, sociological, psychological, and mysti-cal foundations that support a people. Ritual can enact or evoke all or some of a founding mythology. "By participating in a ritual, you are participating in a myth" (Campbell and Moyers *Ep. 3*). Mythologies that guided and shaped mar-tial cultures were heavily imbued with ritual and symbolism. As Western socie-ties moved away from collectivism and toward individualism, the use of intense ritual practice has declined.

A ritual is a narrowly prescribed activity that intends to affect a neurological and physiological transformation in the participant(s). Depending on the readi-ness of the participant, the ritual also works to break past the profane and place the person in touch with the transcendent. A ritual can last a day, several days, weeks, or even months. Many seminal events within martial culture have all the markings of ritual with none of the conscious recognition of the sacredness of these events. As an example, US basic training (boot camp) is a ritual threshold experience that can last three to seven months. Boot camps contain ancient prac-tices of similar components that make up the sacred ritual elements such as lack of sleep, little food, extended days of physical and mental exertion, and rhythmic singing and movement such as synchronized walking and running. All of these practices aid in transformation. Yet the sacredness is not recognized consciously and therefore the inner experiences of the individual are trapped within an industrialized system that officially emphasizes secular logos over spiritual mythos.

US martial culture members are unconscious practitioners of daily lived ritual which coincides with what Stephenson, Eliade, Meade, and others cite as prime examples of authentic ritual (Stephenson 16–19, 63; Eliade Rites and Symbols xx, 1). However, the rituals are hidden by the oft-used term "customs"[4] and, in many cases, are hollow—devoid of life and meaning—because they have been severed from their ancient mythic roots. Hollow though the rituals may be they still retain power. Once the servicemember leaves the military organization and is thrust into the individualistic cultures of the modern world, they are cut off from participating or enacting ritual. This excommunication from something servicemembers-turned-veterans are not even aware they practiced fractures their continued existence as a member of martial culture. The insistence on excising martial cultural norms from the individual is ironic, given how much humanists have bemoaned the ruthless stomping out of Indigenous peoples and their rituals and mythologies. The current emphasis on accepting and celebrating diversity of culture within the US should include martial cultures and emphasize cultural pluralism.

Because martial collectivism is tribally structured, the rituals that bind a martial cultural group together are based on the desire to answer the needs of the individual so as to better provide for the safety and continuity of the larger society. Today, the US military is an organization that fosters martial culture through secular ritual practices. Within US martial culture, ritual survives disguised as custom, tradition, ceremony, or even as scientific application of psychology and physiological conditioning.

Discoveries in the function of the brain-body system indicate that ritual (especially long rituals with numerous repetitions or extreme physiological stress) builds and shapes the structure of the brain during the lifespan of an individual (Hobson et al. 260–84; Dornan 25–36; Graybiel 359–87). Embodied experiences can and do reshape neural pathways through repetitive motion and intense experience. Discoveries that "experiences can modify cortical organization" and brain "plasticity in response to experience and can form synapses in hours and possibly even minutes after some experiences" (Kolb and Gibb) demonstrate that external stimuli, especially an intense experience, can alter and remap the physicality of the brain. Science is now demonstrating the "how" of what ancients understood regarding the effectiveness and value of ritual. The various training regimens, specifically beginning with Basic Training as the first liminal experience of extended trials consisting of multiple rituals, fulfill the conditions required to engage the brain's plasticity and begin remapping neurological pathways. Further, neurological studies show that "the same experience can alter the brain differently at different ages" (Kolb and Gibb).

Today it is understood that the prefrontal cortex is not fully formed until the mid-20s. The discovery that the brain can reconfigure itself throughout the lifespan of a human's existence has great implications (Begley 110–30). External sensory stimuli, if afforded sufficient focus, can remap neurons, cause new neural pathways, and reconfigure the brain-body system so that the individual will act and behave in a way that is conducive to its environment (Siegel 196–97).

Rituals can link the biological survival function to social inclusion, thereby providing the external stimulus linking survival with acceptable values, beliefs, and norms of behavior for a culture. If the participant in the ritual voluntarily accepts the rites, the rites are of a sufficient intensity, and participation incorporates physical exercise, then brain will shape and strengthen new pathways as well as increase neurogenesis within the hippocampus region of the brain thereby generating new neurons to increase speed and connectivity (Begley 49–72; Kolb and Gibb). Repetitive, intense activity that inspires neurogenesis was demonstrated in studies of London taxi drivers in which "years of navigation experience" had resulted in "right posterior gray matter volume increasing and anterior volume decreasing with more navigation experience" (Maguire et al.). The brain rebuilds itself. Martial cultures combine physical, mental, and emotional activities on a daily basis which place the individual servicemember in a state for what neuroscientists are finding are optimal conditions for restructuring the brain through the body.

Martial culture is an embodied experience. Emphasis on locomotion such as running, swimming, or carrying heavy loads builds and expands the capacity of what is sometimes referred to as the combat chassis or the organic body. The application of physical stress and stressors expands consciousness and breaks through the illusion of either self-imposed or externally-imposed limitations, to transform "I can't" into "I can." The utilization of physical stress to create psychological and spiritual expansion is well documented in many religious and mythological practices. Examples range from Buddhist monks using stress positions for hours on end to the ritual that occurs in South America which requires the participant to carry a full-sized Christian cross, and even re-enact crucifixion. Each physical movement can become or trigger a conflict within that is psychological and spiritual, challenging the initiate's preconceptions of the capabilities of the self throughout all spheres of existence. This challenge is intentional; individuals who experience discomfort through exposure to weather and rough terrain yet accomplish difficult deeds learn more about their potential as humans.

The rites US martial culture uses today in training and the experiences of those servicemembers act to restructure the brain and body over the course of years. As psychologist Bessel Van Der Kolk credits the work of Ed Tronick, the brain is a "cultural organ" and the importance of experience as a way of mapping cultural norms into the neurology cannot be overstated in relation to the acculturation process experienced by servicemembers. The duration and intensity of the martial experience make calls for reverse boot camps or other quick methods of integrating servicemembers into civilian cultures unrealistic. The Congressionally mandated Transition Assistance Program focuses on combating veteran homelessness and joblessness by offering programs for job placement and additional education with the goal of job placement. In my experience with this program, the only other aspects that focused on "being a civilian" were a list of "dos and don'ts" regarding how to style hair, speak, stand, and dress. Coincidentally, this list mirrors the subjects that are covered in cross-cultural briefings before servicemembers go to a foreign country. Servicemembers, veterans, and civilians must acknowledge that martial culture is a lifeway that will stay with one indefinitely, and therefore an incorporation of this lifeway while moving forward is far more conducive to a richly holistic life than the suppression and eradication of martial culture within oneself.

A Martial Life Cycle

The graphic below offers a basic visual depiction of a member of the contemporary US martial culture's life cycle which ignores any cultural influence beyond membership to the military organization. While each aspect is expanded in subsequent chapters, it is introduced to provide a basic model (Figure 2.1).

The first major step to joining martial culture is to submit oneself to be tested for qualification at a Military Entrance Processing Center (MEPS). Then they move into a liminal initial training which teaches everything from how to dress, walk, speak, and interact within the new culture, essentially reenacting the social

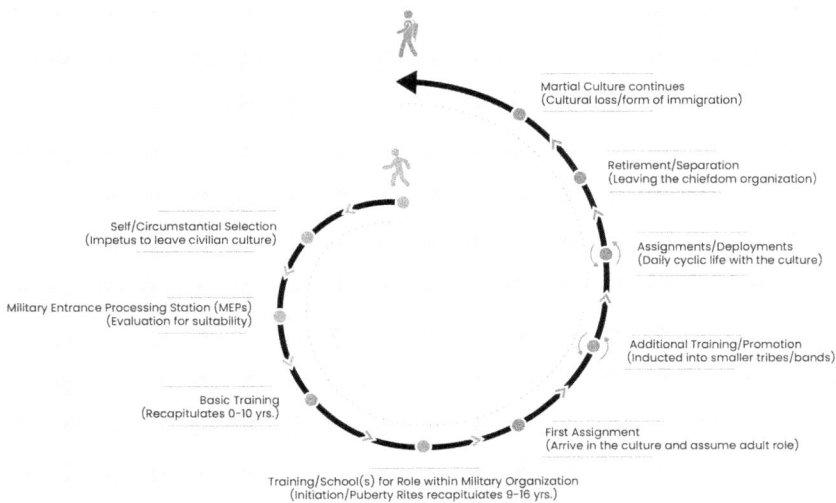

Martial Culture continues
(Cultural loss/form of immigration)

Retirement/Separation
(Leaving the chiefdom organization)

Self/Circumstantial Selection
(Impetus to leave civilian culture)

Assignments/Deployments
(Daily cyclic life with the culture)

Military Entrance Processing Station (MEPs)
(Evaluation for suitability)

Additional Training/Promotion
(Inducted into smaller tribes/bands)

Basic Training
(Recapitulates 0–10 yrs.)

First Assignment
(Arrive in the culture and assume adult role)

Training/School(s) for Role within Military Organization
(Initiation/Puberty Rites recapitulates 9–16 yrs.)

Figure 2.1 A Martial Culture Lifecycle.

Source: Created by the author.

education of age 0–8. From there the servicemembers move to yet another liminal space where they learn their adult role (job) in the military organization
and further sink into martial culture, reenacting ages 9–16 years and serving as
an equivalent to puberty rites in older cultures. Then they move fully into the
military organization and martial culture where everything seems familiar yet
foreign. They begin their existence as a new being and live there, finding themselves moving into different tribes or bands as they attain new skills or move
from one unit to another. Their daily life is much like other martial cultures who
endured endemic warfare. The servicemembers remain in their installations to
train and conduct the day-to-day life of supporting and living in the community
until the servicemember is called to deploy away from their assigned installation. This may take the form of a combat or humanitarian deployment, a tour of
duty aboard a ship, or other activities such as specialized schools or training.

Finally, either voluntarily or involuntarily they will exit, forever changed
through numerous experiences and rituals that shaped their brains, bodies,
minds, and souls into something other than what they had been. The effects
of the mythic dimensions of martial culture are underestimated and are akin to
coming back from adventure with secret knowledge, many times represented in
myth as gold or another precious material or object. The expectation that one can
go back to being a civilian who has gained nothing during their military service
that is of use in civilian culture usually results in coming out of the forest with
gold that turns to ashes (Campbell and Moyers *Ep. 1*). That is, all of the servicemembers' experiences and learning that could be brought forth into civilian
culture are quite often seen as not applicable in civilian culture and therefore
without worth. If the axiom "A person is the sum total of their experiences" is to
be believed, and if those experiences are deemed of no worth, what is the value
of the veteran within civilian culture?

In interviews with several individuals who have served for a substantial
amount of time, a theme emerges on a concept I refer to as martial maturation. Martial maturation is a process by which the individual must continue to
inculcate their continued growth in martial culture, turning from an ephebe-like
self-aggrandizing "hero" self into an elder of the martial lifeway, reflective and
inclusive in their views. If the martial maturation process is interrupted, such as
serving four years and leaving the military organization, there is a danger of the
veteran being stuck in ephebe status regardless of their biological age. It is vital
to have elders who can continue to inform growth, not of skills such as shooting,
but of mental maturity and emotional regulation that comes from long years of
service mixed with the correct level of introspection and reflection.

Conclusion

A proper martial culture exists as the enactment of living a martial mythos. In
the United States today it exists as an ancestrally-oriented, nomadic, totemic

collection of chiefdoms comprised of several tribes and bands. The martial cultural group maintains its own historic and contemporary personalities as guides to behavior and is heavily imbued with symbolism. The individual exists to support the service and gains self-worth from the same. The loss of membership to the tribe often results in the loss of self-worth. Martial culture within the military is hierarchal, with internal caste-like rules of association where every member of the society is a potential regulator of behavior (including family members) (Meadows et al. xvii–xxix). Finally, the lifestyle is nomadic as depicted by the mandatory relocation every three to four years with deployments and missions in between contributing to a constant state of mobility.

Martial culture is heavily misunderstood due to the use of secular lenses that include prejudices and biases from and toward civilian culture that have added up over millennia. This misunderstanding leads to a failure to recognize or appreciate the differences between martial and civilian cultures. So fully has the secular government employee narrative been accepted by both civilian and martial culture that culture shock was not recognized as an affliction affecting veterans until recently (Smith 103–14). Culture shock can be debilitating even when it is fully and consciously expected. Culture shock that is unexpected and unrecognized can lead to greater issues which could manifest as psychological pathology and potentially even suicide.

Many veterans fall victim to substance abuse and a significant number die by suicide. These types of responses are common when a culture is ripped from the sustaining mythology it relies on for existence. In the case of martial culture, it is the individual who is removed from the tribe within the span of a single day, who loses a centrality that may only exist within the subconscious, but exists, nonetheless. The misuse of substances to cope with the loss of identity is a useless attempt by the member of the martial culture who has been thrust into another culture so very different from the one they either consciously or unconsciously identify with. The great power of the prevailing narrative today is that the martial culture itself is to be blamed as an aberration. In other words, individuals blame the military as at best a necessary evil that alters the individual from their natural state. This informs a great deal of "trauma therapy" as linked to "veteran reintegration." However, this view largely fails to recognize that martial culture is holistic and as such, not everything connected to military service is negative. In fact, the values, beliefs, and norms of daily martial culture, I assert, are the key to rebalancing the veteran, rather than trying to force assimilation into a different culture.

The interior anguish and emptiness that results from a loss of culture can be suppressed or repressed, but only to a point. The servicemember-turned-veteran is now a stranger in a strange land. The failure to recognize martial culture as a life-long phenomenon leads to suppression or a seeking for purpose, which in some cases leads to a non-functional, broken form of martial culture.

The misuse and evolution of certain terms has led to the stereotyping of martial practitioners. The civilian culture, which is now the overwhelming majority in America today, applies a standard of normal that paints martial culture as deviant from the norm. In many corners of America, military service is viewed as barbaric, primitive, or a necessary evil. Civilian culture seeks to "correct" many of the values, beliefs, and norms of a person from the martial culture while simultaneously espousing an increase in tolerance and strength in diversity. Further, this norm is actively pushed in secular narrative as the only moral goal, thereby depicting any deviations as bad or, when codified, illegal. There exists an almost missionizing and/or colonizing quality to the phenomenon of correcting social behavior of other cultures to force adaption to one's own norms through law.

Therefore, an exploration of mythologies of martial cultures must first contend with the preconceptions of the individuals in the culture. Mythologies today are largely viewed as metaphors and are seen as the sole property of academia specializing in the liberal arts and psychologists. The preservation of mythology is a godsend in that scholars have kept the myths alive and found vitality within them. Further, if read through a lens of those inheritors of martial culture, the myths can serve as a way to help connect civilian culture with martial culture. Reconnection to myth is what is missing in servicemember and veteran lives. The myths need to be revitalized and reintroduced as depictions of ancestral knowledge for the current servicemember. The mythologies must not be cast aside as old, outdated stories, but reinvigorated for practical application in how to live a human life as a servant to the larger community.

Notes

1 Several European military units are referred to or incorporate the Germanic term *Jäger*, literally "hunter" in their unit name or role designation.
2 *War* as a noun derives from the Late Old English *wyrre*, Old French *werre*, and the Old High German *werra* which convey the concepts of confusion, discord, strife, and Old Saxon *werran* and as a verb meaning "to bring into confusion or discord" ("war"). Wars original etymology describes an altogether different meaning than how it is commonly applied today. It does not include "kill as many as possible" as Bertrand Russell once asserted (Dennen 4). Within military institutions it encompasses all aspects of a struggle, not just physical violence. The Romans viewed verbal rhetoric and debate, as well as declarations of injustice or demonstrations of military capability, as part of this continuum; this elucidates why the entire pantheon would be involved at various stages of conflict (Dumézil *Archaic Roman Religion* 208–209). In this way, war manifests across all aspects of life. As one travels backward in time the roots of war are more primordial and elemental.
3 Originally military derives from "*militari,* 'pertaining to or befitting soldiers; used, done, or brought about by soldiers,'" which is derived directly from Latin's "*"militaris*" of soldiers or war, of military service, warlike," ("military"). Yet De Vaan sees a connection with the root *mīli-it-* "It is tempting to connect *mīlia* [pl.] 'thousand(s)', hence **mīli-it-* 'who goes with/by the thousand'" ("military"). This root which would correspond with how the legions were organized by tens also corresponds with the idea of a large assembled group as in Sanskrit "*melah* "assembly," [and the] Greek *homilos* "assembled crowd, throng" ("military").

4 The US military uses terms such as custom, ceremony, tradition, etc. to describe many practices that would be recognized as ritual. Therefore, when addressing these in this book, I will use the term ritual.

References

Ahern, Jennifer, et al. "'The Challenges of Afghanistan and Iraq Veterans' Transition from Military to Civilian Life and Approaches to Reconnection." *PloS one* vol. 10, no. 7, 1 Jul. 2015, e0128599, https://doi.org/10.1371/journal.pone.0128599

"American Culture: Study in the USA." International Student, www.internationalstudent.com/study_usa/way-of-life/american-culture/

Batalova, Jeanne, and Jie Zong. "Immigrant Veterans in the United States." *Migrationpolicy.org*, 2 Feb. 2021, www.migrationpolicy.org/article/immigrant-veterans-united-states-2018

Beaver, Kevin, et al. "Enlisting in the Military: The Influential Role of Genetic Factors." *SAGE Open*, vol. 5, no. 2, 2015. https://doi.org/10.1177/2158244015573352

Begley, Sharon. *Train Your Mind, Change Your Brain: How a New Science Reveals Our Extraordinary Potential to Transform Ourselves*. Reprint ed., Ballantine, 2007.

"Best for Vets Places to Live 2019 Medium Cities: Charts: Rebootcamp." *Charts*, https://charts.militarytimes.com/chart/7

Bjorkman, Eileen. "We're Still Arguing over Women in the Military?" *Defense One*, 12 June 2023, www.defenseone.com/ideas/2023/06/were-still-arguing-over-women-military/387444/

Blackmarr, Jennifer. Personal Interview. 9 July 2023.

Brown, Elaine K., et al. "'A Woman in a Man's World': A Pilot Qualitative Study of Challenges Faced by Women Veterans During and after Deployment." *Journal of Trauma & Dissociation: The Official Journal of the International Society for the Study of Dissociation (ISSD)* vol. 22, no. 2, 2021, pp. 202–19, https://doi.org/10.1080/15299732.2020.1869068

Brunt, Charles D. "Two Navajo Women May Have Been America's First GI Janes." *Albuquerque Journal*, www.abqjournal.com/887174/were-navajo-women-first-gi-janes.html, accessed 12 Aug. 2020.

"Built to Serve." *Team Rubicon*. teamrubiconusa.org/

Caforio, Giuseppe, et al. *Handbook of the Sociology of the Military*. Springer, 2018.

Campbell, Joseph, and Bill Moyers. *Ep. 1: Joseph Campbell and the Power of Myth – "The Hero's Adventure."* 27 Aug. 2018, https://billmoyers.com/content/ep-1-joseph-campbell-and-the-power-of-myth-the-hero%E2%80%99s-adventure-audio/

———. *Ep. 3: Joseph Campbell and the Power of Myth – 'The First Storytellers'*. Bill-Moyers.com, 19 Oct. 2020, https://billmoyers.com/content/ep-3-joseph-campbell-and-the-power-of-myth-the-first-storytellers-audio/

Campbell, Joseph, et al. *The Power of Myth*. Doubleday, 1988.

"Cities with the Most Veterans." *Stacker*, https://stacker.com/stories/1483/cities-most-veterans

Coll, Jose, et al. "No One Leaves Unchanged: Insights for Civilian Mental Health Care Professionals Into the Military Experience and Culture." *Social work in Health Care*, vol. 50, 2011, pp. 487–500, https://doi.org/10.1080/00981389.2010.528727.

Cox, Kate, et al. *Understanding Resilience as It Affects the Transition from the UK Armed Forces to Civilian Life*. RAND Corporation, 2018.

Daugherty, Lindsay. *Defining Corporate Culture: How Social Scientists Define Culture, Values and Tradeoffs among Them*. RAND Corporation, 2007, https://www.rand.org/pubs/working_papers/WR499.html

Davis-Kimball, Jeannine, and Mona Behan. *Warrior Women: An Archaeologist's Search for History's Hidden Heroines*. Warner Books, 2003.

Dennen, J. M. G. van der. "On War: Concepts, Definitions, Research Data: A Short Literature Review and Bibliography." *CORE*, 1 Jan. 1970, https://core.ac.uk/display/12857871?recSetID=

Department of Defense. *2015 Demographics: Profile of the Military Community*, 2015, https://download.militaryonesource.mil/12038/MOS/Reports/2015-Demographics-Report.pdf

Dickson, Marcus W., et al. "Societal and Organizational Culture: Connections and a Future Agenda" *The Oxford Handbook of Organizational Climate and Culture, Oxford Library of Psychology*, edited by Benjamin Schneider, and Karen M. Barbera, online edn, Oxford Academic, 4 Aug. 2014, https://doi-org.ezproxy4.library.arizona.edu/10.1093/oxfordhb/9780199860715.013.0015, accessed 18 May 2023.

Dornan, Jennifer L. "Beyond Belief: Religious Experience, Ritual, and Cultural Neuro-Phenomenology in the Interpretation of Past Religious Systems." *Cambridge Archaeological Journal*, vol. 14, no. 1, 2004, pp. 25–36, https://doi.org/10.1017/S0959774304000022

Dumézil, Georges. *Archaic Roman Religion: With an Appendix on the Religion of the Etruscans*. Johns Hopkins University Press, 1996.

Dunbar, Robin I. M. *Human Evolution: Our Brains and Behavior*. Oxford University Press, 2016.

Eliade, Mircea, et al. *Rites and Symbols of Initiation: The Mysteries of Birth and Rebirth*. Spring Publications, 2017.

"Facts on Veterans and Entrepreneurship." Small Business Administration, www.sba.gov/content/facts-veterans-and-entrepreneurship

France, Jason. Personal Interview. 28 June 2023.

Graybiel, Ann M. "Habits, Rituals, and the Evaluative Brain." *Annual Review of Neuroscience*, vol. 31, 2008, pp. 359–87, https://doi.org/10.1146/annurev.neuro.29.051605.112851

Hanson, Victor Davis. "War and History, Ancient and Modern." *Uncommon Knowledge*. March 17, 2010. Hoover Institution. Feb. 18, 2018, https://www.youtube.com/watch?v=3VdaOUBoZ3E

Hobson, Nicholas M., et al. "The Psychology of Rituals: An Integrative Review and Process-Based Framework." *Personality and Social Psychology Review*, 2017, https://faculty.haas.berkeley.edu/jschroeder/Publications/Hobson%20et%20al%20Psychology%20of%20Rituals.pdf

"Home." *Boulder Crest Foundation*, https://bouldercrest.org/

Johnson, Travon S., et al. "Service Members Prefer a Psychotherapist Who Is a Veteran." *Frontiers in Psychology*, vol. 9, no. 1068, 2018, https://www.frontiersin.org/article/10.3389/fpsyg.2018.01068

Jones, David E. *Women Warriors: A History*. Potomac Books, 2005.

Kuhn, Steven L., and Mary C. Stiner. "What's a Mother to Do? The Division of Labor among Neandertals and Modern Humans in Eurasia." *Current Anthropology*, vol. 47, no. 6, 2006, pp. 953–81. JSTOR, www.jstor.org/stable/10.1086/507197. Accessed 12 Oct. 2020.

LeBlanc, Steven A., and Katherine E. Register. *Constant Battles: The Myth of the Peaceful, Noble Savage*. St. Martin's Griffin, 2003.

Lévi-Strauss, Claude. *Structural Anthropology*. Basic Books, 1963.

Little, Anthony C., et al. "Social Learning and Human Mate Preferences: A Potential Mechanism for Generating and Maintaining between-Population Diversity in Attraction." *Philosophical Transactions of the Royal Society of London. Series B, Biological Sciences*, vol. 366, no.1563, 2011, pp. 366–75, https://doi.org/10.1098/rstb.2010.0192

Lyon, Bruce E., and Robert Montgomerie. "Sexual Selection Is a Form of Social Selection." *Philosophical Transactions of the Royal Society of London. Series B, Biological Sciences*, vol. 367, no. 1600, 2012, pp. 2266–73, https://doi.org/10.1098/rstb.2012.0012

Maguire, Eleanor A., et al. "London Taxi Drivers and Bus Drivers: A Structural MRI and Neuropsychological Analysis." *Hippocampus*, vol. 16, no. 12, 2006, pp. 1091–101.

Mansoor, Peter R., and Williamson Murray. *The Culture of Military Organizations*. Cambridge University Press, 2019.

Markel, Nicholas, et al. "Resiliency and Retention in Veterans Returning to College: Results of a Pilot Study" *Paper presented at the Veterans in Higher Education Conference: Listening, Responding, Changing for Student Success*, Tucson, AZ. September 2010 University of Arizona and Southern Arizona VA Health Care.

Martin, Carol Lynn, and Diane N. Ruble. "Patterns of Gender Development." *Annual Review of Psychology*, vol. 61, 2010, pp. 353–81, https://doi.org/10.1146/annurev.psych.093008.100511

McCaslin, S. E., et al. "Military Acculturation and Readjustment to the Civilian Context." *Psychological Trauma: Theory, Research, Practice, and Policy*, vol. 13, no. 6, 2021, pp. 611–20, https://doi.org/10.1037/tra0000999

McCoy, Brady. Personal Interview. 6 June 2023.

Meadows, Sarah O., et al. "The Deployment Life Study: Longitudinal Analysis of Military Families across the Deployment Cycle." *RAND Corporation*, 2016, https://www.rand.org/pubs/research_reports/RR1388.html

Meadows, William C. *Kiowa Military Societies: Ethnohistory and Ritual*. University of Oklahoma Press, 2010.

Meredith, Lisa, et al. "Identifying Promising Approaches to U.S. Army Institutional Change: A Review of the Literature on Organizational Culture and Climate." *RAND Corporation*, 2017, https://www.rand.org/content/dam/rand/pubs/research_reports/RR1500/RR1588/RAND_RR1588.pdf

Meyer, Leisa D. *Creating GI Jane: Sexuality and Power in the Women's Army Corps during World War 2*. Columbia UP, 1996.

"Military, adj. and n." *OED Online*, Oxford University Press, September 2020, www.oed.com/view/Entry/118428, accessed 5 November 2020.

"Mission." Team RWB, 10 Dec. 2020, www.teamrwb.org/about-us/mission/.

Moskos, Charles C., J. R. "The All-Volunteer Military: Calling, Profession, or Occupation?" *Parameters*, vol. 40, no. 4, 2011, pp. 23–31, *ProQuest*, https://search-proquest-com.ezproxy2.apus.edu/docview/867412592?accountid=8289

Okafor, Oliver Nnamdi and Kenneth Kalu. "Integration Challenges, Immigrant Characteristics and Career Satisfaction for Immigrants in the Field of Accounting and Finance: An Empirical Evidence from Canada." *Critical Perspectives on Accounting*, 2023, https://doi.org/10.1016/j.cpa.2023.102602

Olenick, Maria, et al. "US Veterans and Their Unique Issues: Enhancing Health Care Professional Awareness." *Advances in Medical Education and Practice*, vol. 6, 1 Dec. 2015, pp. 635–39, https://doi.org/10.2147/AMEP.S89479

Osuna, Freddy. Personal interview. 14 June 2023.

Parker, Kim, et al. "The American Veteran Experience and the Post-9/11 Generation." Pew Research Center, September 10, 2019, https://www.pewsocialtrends.org/wp-content/uploads/sites/3/2019/09/PSDT.10.09.19_veteransexperiences_full.report.pdf

Ricks, Thomas E. "The Great Society in Camouflage." *The Atlantic*, Atlantic Media Company, 25 Sept. 2014. www.theatlantic.com/magazine/archive/1996/12/the-great-society-in-camouflage/376732/.

Rodriguez, Riccardo. Personal interview. 16 June 2023.

Sapolsky, Robert M. *Behave: The Biology of Humans at Our Best and Worst*. Penguin Books, 2018.

Schafer, Amy. "Generations of War: The Rise of the Warrior Caste and the All-Volunteer Force." *Center for a New American Security (En-US)*, 2017, www.cnas.org/publications/reports/generations-of-war.

Schmidt, H. "'Hero-Worship' or 'Manipulative and Oversimplifying': How America's Current and Former Military Service Members Perceive Military-Related News Reporting." *Journal of Veterans Studies*, vol. 6, no. 1, 2020, pp. 13–24, https://doi.org/10.21061/jvs.v6i1.156

Schmookler, Andrew. *The Parable of the Tribes: The Problem of Power in Social Evolution*. University of California Press, 1984.

Schulte, Brigid. "Women Soldiers Fought, Bled and Died in the Civil War, then Were Forgotten." *The Washington Post*, 29 Apr. 2013, https://www.washingtonpost.com/local/women-soldiers-fought-bled-and-died-in-the-civil-war-then-were-forgotten/2013/04/26/fa722dba-a1a2-11e2-82bc-511538ae90a4_story.html

Semczuk, Nina. "3 Military Behaviors That Don't Work in the Civilian Workplace." *Military.com*, 18 Sept. 2018, https://www.military.com/veteran-jobs/career-advice/3-military-behaviors-dont-work-civilian-workplace.html

Shepherd, Steven, et al. "The Challenges of Military Veterans in Their Transition to the Workplace: A Call for Integrating Basic and Applied Psychological Science." *Perspectives on Psychological Science: A Journal of the Association for Psychological Science*, vol. 16, no. 3, 2021, pp. 590–613, https://doi.org/10.1177/1745691620953096

Siegel, Daniel J. *The Developing Mind: How Relationships and the Brain Interact to Shape Who We Are*. Guilford Press, 2015.

Skiles, Zach. Personal Interview. 25 July 2023.

Smith, M. J. "Can Veterans Experience Acculturative Stress?" *Journal of Veterans Studies*, vol. 9, no. 1, 2023, pp. 103–14, https://doi.org/10.21061/jvs.v9i1.371

Stephenson, Barry. *Ritual: A Very Short Introduction*. Oxford UP, 2015.

Street, Sally E., et al. "Human Mate-Choice Copying Is Domain-General Social Learning." *Scientific Reports* vol. 8, no. 1715, 29 Jan. 2018, https://doi.org/10.1038/s41598-018-19770-8

Theokas, Christina. "Shut Out of the Military: Today's High School Education Doesn't Mean You're Ready for Today's Army." *The Education Trust*, 10 June 2019, https://edtrust.org/resource/shut-out-of-the-military-todays-high-school-education-doesnt-mean-youre-ready-for-todays-army/

Tick, Edward. "Our Mission." *Soldiersheart.* www.soldiersheart.net/copy-of-home.

Toler, Pamela D. *Women Warriors: An Unexpected History.* Penguin, 2019.

Treadwell, Mattie E. *United States Army in World War II, Special Studies, the Women's Army Corps.* Defense Technical Information Center, 1971.

Voss, Nathan and James Ryseff. "Comparing the Organizational Cultures of the Department of Defense and Silicon Valley." *RAND Corporation*, 2022, http://www.rand.org/t/RRA1498-2

Walsh, Michele E., et al. "Unpacking cultural factors in adaptation to type 2 diabetes mellitus." *Medical care*, vol. 40, no. 1, 2002, pp. I129–39, https://doi.org/10.1097/00005650-200201001-00014

"War and Sacrifice in the Post-9/11 Era." *Pew Research Center's Social & Demographic Trends Project*, 30 May 2020, www.pewsocialtrends.org/2011/10/05/war-and-sacrifice-in-the-post-911-era/

"War, n. 1." *OED Online*, Oxford University Press, September 2020, www.oed.com/view/Entry/225589. Accessed 5 November 2020

Waters, Mary C., and Marisa Gerstein Pineau. *The Integration of Immigrants into American Society.* National Academy Press, 2015.

3 Navajo

The *Enemy Way* ceremony is based on the origin mythology of the Navajo, and consists of long periods of purification, powerful sand paintings meticulously created with colors of power, and singing the history of the *Diné* (Navajo) people.[1] This chapter focuses on the *Enemy Way* ceremony as a living mythology whose story of the Two brothers, the original warriors of the *Diné* and sometimes referred to as War Gods, is unknowingly lived by US servicemembers today. US martial culture, already owing foundational cultural inheritance to Native American peoples, should look to their ceremonies and rituals. Particularly those associated with the warrior tradition, as ways of learning how to develop a more holistic martial lifeway.

US martial cultures are older than the nation-state itself, with roots in both American Indigenous Peoples and European warrior traditions. In many ways, the Scotch-Irish peoples who immigrated from the borderland regions of Great Britain, themselves descended from peoples who had fought guerrilla campaigns against organized armies in Britain for nearly 800 years, embraced the Indigenous cultures of the Appalachian region to such a degree that many "went native" upon immigrating to America (101–102, 106 Woodard). Another example of this amalgamation of martial cultures is the introduction of the Eurasian horse. This precipitated a symbiotic relationship between this beautiful creature and many of the Indigenous Nations of the North American plain which rivaled those of the original horse peoples in the martial cultures of the steppes of Eurasia. The intertwining of martial lifeways is symbolized today in the naming of US helicopters for Native American tribal groups, as well as the crossed arrows insignia of current US Special Forces. These are examples of recognition of the influence and debt owed to Native Americans for their foundational and influential role in the development of US martial culture. The original Indigenous martial cultures existed well before European contact and they heavily influenced the formation and continued evolution of the US martial culture. Indigenous peoples have served in the US military in every conflict the United States has engaged in and are themselves founding members of US martial culture.

DOI: 10.4324/9781032613222-3

The traditions of Indigenous American martial cultures have kept alive important rituals related to service. But there is also a recognition that the civilian culture has secularized service, separating civilian culture from the martial lifeway. According to Professor Jeff Means of the University of Wyoming, a member of the Oglala Sioux Tribe and Marine Corps veteran,

> The US, to an extent, ignores that militaristic part of society because it's not what we would consider a larger part of American culture. It has been separated to a tremendous degree. Most people have no idea what military service is like, what combat is like. So therefore, they have no empathy. (Simkins and Barrett)

Means feels that the United States and the Indigenous Nations have completely different worldviews and that the United States largely views the military as outside of American culture. This disconnected relationship between martial culture and the larger society is out of step with how many Indigenous peoples view military service—an intrinsic part of existence. The recognition and acceptance of the warrior dimension of life allows for the integration of effective living rituals that serve and guide the warrior in all stages of their life cycle.

If mythology is rooted in the land and informed by the environment, then the *mythos* of US martial culture should include Indigenous mythologies, which are tied to the land. The land (or home) is, in many ways, one of several reasons why many people voluntarily enter the US martial culture. As D.J. Vanas, a member of the Odawa Nation and an Air Force veteran puts it, "This is our home, it always has been and always will be, and we sign up to defend that" (Simkins and Barrett). The idea of serving not necessarily the United States but the land on which the Indigenous nations live is extremely important (Schilling). They are not the only ones who tend to see their service as an idea of place and people smaller than the geopolitical boundaries of the United States.

Of those who join the US martial culture, some do so for the land and their connection to it. Some will fly flags of their home states or carry symbols of home with them. Usually, servicemembers do not join as a way to serve elected officials or appointed bureaucrats, both of whom change almost constantly. Reasons for serving vary to include everything from family tradition, economic need, higher education, adventure, and even (when there was a draft) volunteering to join a preferred service rather than be drafted, usually as a way of avoiding combat. The desire to serve something outside oneself is usually mixed in with these other motivations. In this way, servicemembers may hold within themselves a vision of an idealized version of their land that they are serving, and an idealized version of the military that exists to serve the land.

The *Diné* mythology is the basis for the *Enemy Way* ceremony which serves as an example of the life cycle journey taken by US servicemembers. Understanding the journey of the Two brothers, the original warriors of the *Diné*, gives the servicemember within the US martial cultures both a guide and a touchstone, rooting them to the land they serve. Further, learning an Indigenous Nation's mythos of warriorhood may spur the acceptance of other Native mythologies which speak to all within US martial culture.

While there are countless Indigenous Nations ceremonies, the *Diné Enemy Way* ceremony is explored because there are several text versions available. Additionally, this particular ritual is for those who are or have served in martial culture. This myth lays bare the necessity of ritual practice for the well-being of martial practitioners. This chapter explores the ceremony for relevance to US martial culture through two lenses, mythological and martial, while respecting that the author is not of Indigenous Nations descent and as such cannot speak for those who are.

The ceremony is based on a portion of the *Diné* origin mythology. There have been several recorded versions of the myth. The first, and perhaps most detailed recording, was made by a military surgeon named Washington Matthews who made observations of the *Diné* and their ceremonies from 1880 to 1895.[2] Hasteen Klah's version of the origin myth, related to Mr. and Mrs. A. J. Newcomb and Mary Wheelright, was published in 1947. Lastly, Jeff King, a *Diné* medicine man and former scout for the US Army, recounted the myth to Maud Oakes in 1943 who published it under the title, *Where the Two Came to Their Father,* focusing solely on the *Enemy Way* ceremony.

The Two War Gods' journey is representative of nearly every stage of life a modern member of the US martial culture goes through. The *Enemy Way* ceremony connects organically with martial experiences of all persons who serve in the US military today and should be incorporated into the whole of US martial culture.

"This story […] goes everywhere, because it is full of power—good power. My story has no evil in it. It is straight. It is to save and protect people. It is to save and protect men going to war, or in enemy country" (King et al. 15). A vital part of the story is the retelling of the cosmological beginning, reminding those who participate in the ceremony of the *Diné's* mythic history. The myth connects those who will move into the wilderness with the source of the cosmos.

The ceremony is typically performed at least twice, once as the initiate is about to depart and again upon their return. The ritual and the sand paintings remain the same, yet the participant has changed and is a different creature than they were before the first ritual. Therefore, the ceremony itself, if the participant is open to the experience, is experienced anew and initiates a different internal response.

The *Enemy Way* begins with the story of a need brought about by the *Diné* being attacked by monsters from the wilderness (King et al. 26). The *Diné* Creation myth describes the monster as an entity with excessive "characteristics of alterity; […which causes] an ethnocentric reaction against what [is] deemed 'different' and 'alien'" (Kearney 41). Monsters are set apart from and identified with conflict, usually by the people who excluded them.

In the myth, the monsters are children of the *Diné* who have been abandoned in the wilderness. These children survive the travails of such an existence and return to the *Diné* as destroyers. A choice must be made to either allow the continued attacks on the people by the monsters or create the Two, and bestow on them the power to defend the *Diné* by directly engaging with the monstrous other.

The Myth

A man has a dream; a premonition of the intention of the monsters to return one day. However, despite this gift of foresight, the people's prayers and chants are not enough to stall the coming calamities. The monsters begin to prey on both the *Diné* and the Pueblo peoples. Big Giant, whose home is Turquoise Mountain, also known as Mount Taylor, leads the monsters.

The monsters hunt and consume the people until only four remain. The four who constitute what remains of the *Diné* retreat to *Tse'lakaiia* (White Standing Rock), where they find a turquoise statue of a woman. Talking God comes to them and tells them that in 12 nights time, they must bring the statue to the top of *Tsolihi*.

On the appointed day, the *Diné* climb the sacred trail up the mountain and find a pantheon of sacred beings awaiting their arrival near the summit. A buckskin is laid down, and the turquoise statue is laid on it and then is covered by another buckskin. By the power of the gods, the statue transforms into a woman. She is named Estsanatlehi, or Changing Woman. Upon completion of the ceremony, all depart save Changing Woman. Changing Woman proceeds to the mountain's peak, finding herself alone save the sun and a small waterfall below.

The mother of The Two, known as Changing Woman, is the child of Darkness and Dawn. Because of this parentage, she holds the tension between the worlds of light and darkness and can distinguish between the two. This is why, in her eternally liminal state, Changing Woman can advise the *Diné* when the wisdom of discernment is required. She serves as a sighted Lady Justice, a source of balance. When the world of the *Diné* is so tipped in favor of monsters, it is only natural Changing Woman mother the two children who will provide the ballast necessary to rebalance the world.

Changing Woman is visited by the Sun and four days later she births a baby. Her second conception comes upon her after she bathes in a certain pool under the small waterfall. Four days later she gives birth to another son. The children, remaining nameless until their father the Sun bestows their names, are referred to as the Two or the two brothers.

The early years of the Two describe the training and strengthening that must take place through the involvement of the gods in the rearing of the Two. The first of the three divine interventions in the two boys' rearing happens at the hour of the children's births, as Talking God and Water Sprinkler midwife the children into the world. Talking God takes the boy born of the Sun and washes him. He laughs in delight at the baby and makes slicing gestures, as if cutting the child up and tossing pieces away. Talking God and Water Sprinkler call the two boys grandchildren. They leave Changing Woman and her children, promising to return in four days.

To keep them safe from crows, owls, buzzards, as well as the monsters, Changing Woman digs holes in the earth to hide the two children. The holes act as a second womb, one of the earths itself. Each time she brings them out, they reenact the emergence myth and paint the undeniable view that they are of the land which they are dedicated to defending.

The boys go through cycles of growth in four-day intervals. Their mother creates bows and arrows and the two boys play with these daily. Despite being admonished to stay within sight of their home, the two boys go in the four different directions (East, South, West, and North) encountering Coyote, crow, buzzard, and owl, respectively. The four creatures the boys encounter are spies for four monsters. The children attempt to engage each spy with their arrows but miss each time. Their failures demonstrate that having useful tools does not equate to having the ability to use them effectively. It is the entirety of the self that must be trained and shaped and only then can one competently wield an instrument of power, be it a weapon or a healing balm. These creatures, agents of the monsters, inform their overseers of the children's existence.

Changing Woman understands the purpose behind what the children are destined to do; however, it is her duty to protect the secret of their existence for as long as possible, to give the two boys time to mature enough to be ready for their destiny. The rash behavior of the boys that leads to their discovery by the spies causes consternation for their mother. Changing Woman tells her children that the monsters know they exist and will now come for the two brothers.

The two boys become aware that what was played is now very real. The recognition of mortal danger in the world serves to shape their future existence as warriors. The children recognize the threats to their people and have familiarized themselves with tools that are typically used in defending people by taking life. When the four days have passed, and the Two brothers have reached the age of 12, the gods return. Talking God and Water Sprinkler test the two by challenging them to a race around a nearby mountain. The boys begin to fail in strength before the race is done, and both gods drop back to beat them with sticks, prodding them on. The gods leave, promising to return at the end of another four days. The boys, acknowledging their unpreparedness, begin to train.

After the gods leave, Little Wind tells the boys "the old ones were not such fast runners, after all, and that if the boys would practice during the next four days, they might win the coming race. So for four days they ran hard, many times daily around the neighboring mountain" (Matthews and McNeley 106). Here Little Wind inspires the boys with hope and sustains their drive to constantly push themselves to be better than they were before. The work is not focused on slaying monsters but simply pushing themselves beyond what they were and inexorably beyond the gods themselves.

[W]hen the gods came back again the youths had grown to the full stature of manhood. In the second contest the gods began to flag and fall behind when half way round the mountain, where the others had fallen behind in the first race, and here the boys got behind their elders and scourged the latter to increase their speed. The elder of the boys won this race, and when it was over the gods laughed and clapped their hands, for they were pleased with the spirit and prowess they witnessed. (Matthews and McNeley 106)

The gods are impressed by the heart and spirit of the Two. To overcome the threshold guardian is to defeat the fear within and move forward. Further, to pass through the gate is to take on the attributes and powers previously ascribed only to the threshold guardians.

The boys leave their mother and travel east, seeking the home of the Sun to ask for aid against the monsters. Their mother, Changing Woman, eventually finds their footprints. The elder walks on the right, the younger walks on the left for youth should be on the left and aged wisdom on the right (King et al. 37–38).[3]

By leaving their mother and the gods, the two boys have begun their journey through this liminal space of learning and testing. The first monster encountered is Sand Dune Boy, who dwells in the desert, grabbing travelers and pushing them deep into the sand to kill them. He attempts to grab the two boys and misses. Their counteraction to the attack is not to escalate to violence. They evade the attack and then they "sang songs to the monster and prayed to him; and as he had never been treated like this before, he let them go" (King et al. 37). While all others who had passed that way previously perished, they did not. Instead of giving in to fear, the boys use their compassion (perhaps flattery with a hint of guile) to overcome the conflict of interests between themselves and Sand Dune Boy.

The appeasement of Sand Dune Boy allows the two boys to leap the four mountains and from that moment on they pass into the domain of the transcendent. The Two are referred to as "earth people" and are continually reminded by everyone they encounter that they have passed beyond the earthly realm; a liminal space between earth and the home of the Sun.

Meeting an old woman named Old Age, the Two endure a painful trial of physical disability. By ignoring Old Age's warning and walking on her path, the boys begin to feel "heavy and then they stooped and their steps grew shorter. They became bent [...] Finally, they could not walk at all" (King et al. 38). They have become old because they fail to listen to Old Age and heed the wisdom of one who has walked the path many times.

The two boys are learning to listen to those they encounter. Old Age returns them to their youthful bodies, admonishing them to obey her words. The boys continue their journey, having learned that they need to listen to people who are wiser than they are. With this lesson learned, they meet their next teacher, Spider Woman.

Spider Woman is kind "but she had something besides kindness within herself" (King et al. 39). This something is the stern evaluation of their actions and attitudes as they interact with her. She takes measure of them for she "knew what they were thinking" (King et al. 39). Spider Woman equips the children for the next stage of their journey, granting them strength and knowledge to pass the final tests the boys will encounter. She instructs them to eat four baskets of powerful foods and fortifies them with the power unique to each. The older brother is given a piece of turquoise to swallow and the younger brother is given a white shell to swallow "to make the heart strong and to give them courage" (King et al. 40). Additionally, each is given a live eagle feather, stolen from the Sun. The eagle feathers are meant to protect them and offer aid when they are in peril. Four more obstacles are encountered, yet as the boys are still learning, they are allowed to begin each trial and find themselves without an answer; help comes, usually in the form of Little Wind, the one who aided in them in their race against the gods when they were younger.

The final tests and advanced lessons are received at the house of the Sun which is guarded by a series of supernatural beings. From here on, the two boys are now referred to as men. They have completed all but the last training, grown to adulthood with one final series of trials in the house of their father the Sun.[4] The Two are able to enter unmolested where they are aided by children of the Sun who wrap them in clouds to hide because, "father is away, but when he returns he will kill you, or hurt you, for he lets no one in" (Matthews and McNeley 111; King et al. 41). The Sun is the last terrible threshold guardian who keeps the knowledge and instruments necessary for the two to begin their service to the *Diné*.

The Sun searches through the house and eventually finds the two boys who are now men. In order to prove they are not his sons; the Sun contrives a series of tests. These tests serve as a way of guarding the final knowledge from those who are not worthy of receiving it. All of the previous instructions, journeys, and even the mistakes and seemingly chance encounters, prove absolutely necessary for the Two to pass the Sun's tests, which function as threshold guardians.

The first challenge is a sweat lodge. The daughter of the Sun, who hid the men initially, comes to their aid once again by digging two holes within the floor of the sweathouse and covering the opening with rocks to camouflage them from the Sun. This is exactly how Changing Woman protected the boys as babes from the monsters after their births. The correlation between the Sun who, though he may be the one who will eventually give the men all they require in both physical instruments and sacred knowledge, initially seems indistinguishable from the monsters that represent death.

The Sun's daughter tells the Two about the holes she has prepared.[5] Then, the Sun tells the Two to undress and enter the sweat house. They do so, though "it was so hot they could hardly enter" (King et al. 43). Unfortunately, the feathers they had been given for protection against all dangers are in their clothing and here, divested of all coverings, the Two enter the dark forge of the Sun. The entrance is covered and the Two quickly find the holes and place the rock coverings over them, again reentering the earth as both tomb and womb. Water is thrown on the hot rocks and irons within, causing the rocks to pop, hiss, and run around thereby creating a great cacophony and emitting vast quantities of steam. As the sounds quiet and the heat storm passes, the Two emerge from their holes and hide the openings with rocks. The Sun, surprised that the Two did not die, bids them to follow him to his house.

On the way, the Two receive more surreptitious assistance on how to survive the next test. The Sun takes them to a room with a platform above four poles covered with flint knives. Little Wind whispers to the Two to have their eagle feathers close. Each man was thrown down twice, both times missing the poles and landing easily on their feet.

The Sun shows them three rooms, one to the East, one to the South, and one to the West, all filled with various useful things. Each time, the Sun asks, "Is this what you want?" to which they reply, "We shall need this in the future, but not now" (King et al. 44). The Sun then takes them into a room where all manner of great weapons are stored and offers them to the Two. The brothers say they do not need them but, on the advice of Little Wind, ask for a small bundle over the doorway. "In the bundle was a very powerful medicine, which would protect them when they killed the monsters […] and by this medicine Sun would know when they were in danger and needed help. It was for their protection" (King et al. 45).

This event signals the end of the first tests, and the Two are taken to the Sun's daughter. She puts each man on a type of rug and applies balms to their bodies. Then, she begins to knead each of the brothers, changing their bodies and giving them supernatural strength and features. They are imbued with the physical forms of the Sun's children; their outward appearance reflects their true inner selves. "This is the second birth through the father" (Campbell *Transformations of Myth* 42). They enter the Sun's house with their mother(s) features and the bodies given them by the Mother, yet nameless. They now possess characteristics and bodies shaped by both the Mother and the Father. This completion of their selves allows for the bestowing of names and powers associated with them.

The Sun begins the investiture process of the Two. To the older, the Sun gives a man-shaped piece of jet rock to swallow, and an identically shaped piece of turquoise to the younger to swallow. The Two are invested with their armor; the older dresses in "black flint armor, and the younger in blue" (King et al. 46). The older is named Monster Slayer and the Sun calls him son. The younger is named Child Born of Water, and the Sun calls him grandson. Sun instructs them on how to kill Big Giant, the leader of the monsters, even though Big Giant is also the son of Sun.

The Two asking for the Sun's assistance in killing Big Giant creates a moment of crisis in the myth. For the Sun also "claimed the giants as his children" and therefore initially recoils at the idea of the Two wanting to kill Big Giant (Klah and Wheelwright 84). However, upon hearing what the monsters are doing on the earth to the *Diné*, he acquiesces and "taught them how to wear the armor, and how to shoot the lightning arrows, and how to use the stone knife, the big hail, the cyclones, and also how to use the magic kehtahn [prayer stick]" (Klah and Wheelwright et al. 84).

As the Two are preparing to leave, they are joined by two more, Reared Underground and Changing Grandchild. Sun instructs the four, "When you go down to earth and start killing the monsters, you must not do it without my permission. And after you kill all the monsters, I will help you again. Go outside and stand facing the east, in single file." (King et al. 46). The Two are now Four: Monster Slayer, Child Born of Water, Reared Underground, and Changing Grandchild.[6]

The Two brothers are given one final test, to correctly identify each of the four mountains. Upon successfully completing the test they are bestowed with all of the Sun's knowledge and a feather for each given by the Sun. The Four then descend to the earth, the Two brothers, Monster Slayer and Child Born of Water, alight on Turquoise Mountain, and Reared Underground and Changing Grandchild are conveyed elsewhere.[7]

The Two touch down on to the earth, having passed their trials, to fulfill their destinies. This is the time of "the new May Moon [when] all that grows has its full strength" (King et al. 47). They descended "but did not know it till they had landed on Mount Taylor [Turquoise Mountain], for clouds and fog were all about them [… and] the May moon was shining" (King et al. 47).

"After [Monster Slayer and Child Born of Water] had landed, the dawn broke, and they put on their armor; and Sun gave them lightning arrows, spears, and weapons" (King et al. 47). The Two now move forward, having survived their initiation rituals to get to the Sun as well as the tests the Sun put before them to prove their worthiness of his bloodline. They have moved from looking to the authority figure of the Sun for aid and instead, have become both aware of, and competent with, their own powers.

The Two made their way to the Hot Spring to test their weapons (King et al. 47). "Big Giant […] saw them trying out their lightning arrows; he saw them reflected in the lake" (King et al. 47). The Two, in turn, saw Big Giant also had armor, "What a big man he is! Where shall we aim, to kill him?" (King et al. 47). Big Giant, confusing the Two with their reflections in the lake, drank the entire lake and not seeing them in the mud, thought "he had swallowed them [and so] threw the water up" (King et al. 47). He did this a total of four times, which sapped him of strength. Laying on his back, he glimpsed the Two "outlined against the sky" and realized he had been fooled by their reflections. Sun observed, from high above, the opening exchanges (King et al. 47).

Little Wind, continuing to act as the advisor, orchestrates the battle and informs the Two when and how to engage Big Giant, "He is getting ready to shoot. Stand on your feathers, and it will lift you; and when the arrow passes under, run quickly and get it" (King et al. 48). In this way, they maneuvered around the arrow shots of Big Giant and collected the four arrows. Little Wind said, "Let Sun shoot at [Big Giant]first" (King et al. 48). Sun shoots two lightning arrows down on Big Giant from above, weakening but not killing him. The older brother, Monster Slayer, then fires a "lightning arrow [and hits Big Giant] in the heart from one side" and the younger brother, Child Born of Water shoots Big Giant "through the heart from the other side" (King et al. 48). The Two pull their flint knives and rush Big Giant and "put their knives through his heart" (King et al. 48) and, after a series of death throes, Big Giant finally succumbs and dies.

Once Big Giant is killed, there is still the danger of him coming back to life and continuing to fight. "When the giant was dead, the blood ran." Little Wind said, "If the blood runs up Mount Taylor, he will come to life. Put a flint knife in front of it, so that it can't flow" (King et al. 48). The Two have to do the same to keep the blood from flowing toward any of the other three sacred mountains as well as to protect the water. "Little Wind had told them to take [Big Giant's cap] and destroy it, because if it were left, it might cause sickness in the future" (King et al. 48). After the battle with Big Giant, the Two go home to their mother. On the way "they met Talking God, and they embraced." He said, "I am happy you have killed the giant, and that no harm has come to you. I will sing and […] as I sing you will have more power" (King et al. 49). Additionally, he gives them 6 songs to sing that would give them strength and eventually 12 songs (later known as the Talking God Songs) came into being. The Two gain strength from the songs which "carry" them from Turquoise Mountain all the way home.

When they arrive at their mother's house "there were no footsteps, and no one was there" (King et al. 49). They find that their mother and the *Diné* are hiding from the monsters. "When they had found her, they told her of their adventures. They told her that they had killed Big Giant. Their mother was proud of [The Two], and the people also" (King et al. 49). Yet there was little respite. The Lesser Monsters still threaten the people, and so the Two brothers begin a cycle consisting of brief moments at home between forays into the wilds seeking the monsters.

The Two, informed and aided by Little Wind, begin the cycle required by those who participate in endemic conflict. The Two are home, receive word of a monster that threatens the *Diné*, leave to defeat the monster, and return to tell "their mother what they had done" (King et al. 51). The Two move into the wilderness and confront four monsters: Horned Monster, Monster Eagle, Slayer with Eyes (Snake), and Bear That Tracks. These encounters all resulted in the death of the monster and, in the last three cases, the death and dismemberment of the great monsters leads to the origin story of the eagles, snakes, and bears. These encounters are described in detail, with challenges involving stealth, subterfuge, courage, and aid from those who witness the battles. These are the four animal-like monsters. The fifth and final monster is different.

The final monster was not an animal but a rock. Traveling Stone "would run after people, roll on them and crush them to death" (King et al. 51). Through speed, timing, hiding behind rocks and trees, the brothers made Traveling Stone miss so often that it ran north with the two in pursuit. They fired volleys of knives and Traveling Stone became smaller and smaller. They stopped their pursuit, being warned by Traveling Stone "If you chase me to the north, sickness will come from there" (King et al. 52). The Two settled for driving a much less powerful Traveling Stone from the land of the *Diné*. As King points out, "In some of the other stories, the monsters that were not killed went to the north. From there come sickness, cold, and bad dreams; also, witchcraft from the monsters left over, and bad dreams from the rock" (King et al. 52). Despite the best efforts of the Two, there will always be certain things in the world that must still be overcome in order to protect the *Diné*. This not only includes the animal descendants of the four monsters, but also more elusive foes such as sickness, both of mind and of body.

There are still powers in the world that cannot necessarily be eradicated, but simply combated differently than with the type of force the Two brothers wield. The cycle of endemic conflict takes its toll on the Two brothers, pushing them beyond their limits based on a lifetime of service to the *Diné*.

> When the brothers had returned to Mountain Around Which Moving Was
> Done, they became weak and sick, and each day thinner. The Holy People
> sang and prayed over them, but they still lost weight. They talked it over
> and
>> decided they had killed too much and had gone where earth people should
>> not go. So they moved to Navaho Mountain. There the Holy People
> [reenacted]
> *Where the Two Came to Their Father* [The *Enemy Way* ceremony], four
> times, and they were cured after the fourth time. The Holy People then said
> four prayers, in the four directions, and made the Painting of the Twelve Holy
> People. This gave to them a personal blessing, and came from Blessing Way.
> They then felt fine and could move as before. And they talked of living in the
> future, and of the making of the future people. (King et al. 52)

The ceremony, which is the story of the *Diné* and the story of the Two themselves, is told in its entirety to the brothers in a holy place of liminality. The Two are reminded of why they served, what brought them into existence, their triumphs, and tribulations. It grounds them as children of the land. But one telling is not enough. Four times they must hear it, by the Holy People and on holy ground, where the power of the *Diné* is strongest. There is no condemnation; the *Diné* understand that while the Two "had gone where earth people should not go" yet there is an absolute acceptance that this is what had to be done in order to save the *Diné*. The Two are forever changed yet they neither attract pity nor condemnation for their existence as warriors. They are healed by the continuity of their story and seeing it as a road forward to "living in the future, and of the making of the future people" (King et al. 52).

The Two are moving into a new phase of martial culture as warriors which implies a demonstrative service-ship role in aiding the *Diné* as they move forward into the future. Their bodies will remain as they have been shaped by the children of the sun and their hands, arms, and legs will never forget how to wield the lightning arrows and spears and other weapons. And danger in the world still is prevalent, although they have done much to halt the near genocide of the *Diné* at the hands of the monsters. Further they have the gift of not only taking life but healing it as well. For the first bundle they asked for from the Sun was the medicine they can now teach the people to use.

Martial Culture

Monsters

The "monsters (anaye, alien gods) had been actively pursuing and devouring the people, and […] there were only four persons remaining alive" (Matthews and McNeley 104). The defining of the "other" as "outside" the accepted variance of "normal" in appearance or behavior is an extension of evolutionary behavioral patterns found throughout the animal world (Sapolsky 387–93). Appearance and/or behavior based on values, beliefs, and norms has been a defining marker for distinguishing between same and other peoples.

The monstrous is a common theme throughout the world's myths. Many peoples of the world, like the *Diné*, refer to themselves in their own languages as "The People" which automatically defines who is included and who are "others." The latter are identified in translations of the *Diné* as either monsters or alien gods. Within each aspect of the myth, the approach the Two War Gods take in interacting with each monster is measured in relation to a determined narrative based upon how the world should be uniformly shaped for the *Diné*.

However, within the *Enemy Way* ceremony, monsters are not necessarily evil. The term *evil* implies an infinite, immovable, fixed binary fostering a dualistic relationship to the universe. Cultures like the *Diné* recognize that the binary of good and evil is a false concept and do not include such falsehoods in their interactions with the world. Circumstances arise in which the only possible action is to use force in order to continue to exist. The goal should be to restore balance, not to create ultimate good and destroy ultimate evil. Necessity arises from an impasse between competing needs or wants. An enemy or opponent in conflict, therefore, is not evil but is judged contrary and thus "other" through circumstance. Within the field of this existence, defined by time, there is no such thing as an eternal enemy, for the "eternal is beyond time; the concept of time shuts out eternity" (Campbell and Moyers *Ep. 6*). Realization of the falsehood of binary thinking and the temporality of circumstance is recognized in this myth.

Further, to engage with a monster or alien is to attempt to enter into a dialogue wherein both "us" and "others" learn whether or not

> "a valid sense of selfhood and strangeness might coexist (Kearney 11). For "how could we tell the difference between one kind of other and another— between (a) those [monstrous] aliens and strangers that need our care and hospitality [...] and (b) those others that really do seek to destroy and exterminate" (Kearney 10)?

The *Diné*'s ancient wisdom differentiates between those monsters who can be accepted and included in the world and those who pose a legitimate threat to the existence of the *Diné*.

To enter martial culture is to accept the existence of entities who are or will become enemies while resisting the temptation to "monster" these entities. The rationale for this measured approach is that to demonize (place below) or deify (place above) a current adversary is to fail to understand the opponent. To fall into dehumanizing thinking can make a person underestimate their opponent, thereby losing the battle by not recognizing the full capability of the opponent. Likewise, to deify an opponent will cause a person to overestimate the opponent thereby losing the battle merely from having little hope of success. To avoid this problem, the projections of prejudice and bias informed by fear, anger, loathing, hate, awe, or envy must be recognized and confronted.

Another aspect of monstering as alienation-of-an-"other" is frequently experienced by the servicemember/warrior in the civilian society. Popular cultural entertainment such as films (*Apocalypse Now* and *Platoon*), books, and even academic studies depict martial culture as criminally abhorrent. The US Vietnam veteran was monstered as the nation's scapegoat, holding the sins of America and that false "othering" resulted in denigration of the US servicemember that continued for decades.

Another aspect of monstering is the deification of the monstrous. Portrayal of servicemembers or veterans in popular films such as *Transformers* and *Saving Private Ryan* elevate servicemembers to heroes. The irony is that the perception of another as monster or hero has a great deal to do with personal and collective perceptions of the alien other. Time and experience can change that perceived relationship. A soldier who participated in the invasion of Iraq in 2003 may return in 2006 to train the new Iraqi military which may be composed of the very same individuals the servicemember fought against a scant three years earlier.

The Two brothers of the *Enemy Way* ceremony may themselves be seen as monstrous to the monsters or alien gods they seek to engage. Yet, in the world of activated living that is within the sphere of temporal existence, a person must act from a perceived view of the world as one engages with an infinite number of competing individuals with competing views of the world. The monsters act to disrupt the balance of the *Diné's* world, thus other powers must engage to correct the imbalance.

Call

The creation myth establishes the deities, the *Diné*, as well as other peoples such as the Pueblo, and the monsters. The monsters' attacks on the *Diné* and Pueblo initiate the call for a solution which is answered by the eventual creation of the two brothers. The myth is directly connected to the present when King, in his telling, states, "And [the monsters] would bring their victims, in the evening, for [Big Giant] to eat (just like Hitler)" (King et al. 34). King's allegory of Big Giant as Hitler-like connects the creation myth to King's present and identifies the participant as one of the Two War Gods as well as who the enemy is. As the rise of Big Giant initiates the need for the Two War Gods to come into existence for the *Diné*, the rise of Hitler initiates the need for the young *Diné* who are attending King's ceremony before enlisting. The linking of the mythic age to the now is still very strong at the time King relates his tale to Oakes.

Training

Changing woman gives birth to the first brother and Talking God

> took it aside and washed it. He was glad, and laughed and made ironical motions, as if he were cutting the baby in slices and throwing the slices away. [Talking God and Water Sprinkler] called the children Smali (grandchildren), and they left, promising to return at the end of four days. (Matthews and McNeley 106)

Matthews' description of this moment links the two gods with the role of mentor, a grandparent who has within them the knowledge of how a child must be shaped in order to fulfill the needs of the tribe. Talking God takes measure of the child, delighting in both what he is and will become. Then, Talking God and Water Sprinkler fashion baskets arrayed with immense power and protection and placed the Two brothers within them. This is the role of the grandparent or elder; to pass along the knowledge of the People to those who will move from the dependency of childhood to adolescence when adult roles will be assumed.

The role of Basic Training—or boot camp—is to enact a rebirth and maturation to age ten years. The Drill/Training instructors have the knowledge of the martial culture, and it is their role to act as Changing Woman, Talking God, and Water Sprinkler. They are both parent and grandparent. The instructors will raise the recruit as the Two are raised, first in measuring and then shaping the individual who wishes to be a member of martial culture. To do this, one must first be "cut up."

"Cutting up" begins with the arrival at Basic Training. The recruit-as-initiate will not forget the stormy thunder of the Drill Instructors as they swarm over the newly arrived trainees and can readily recall the confusion such forces of nature produce. The instructors' demeanor begins the process of measuring, cutting up, and discerning the pieces that must remain and those that must be discarded. Air Force retired Senior Master Sergeant Jennifer Blackmarr, a former Basic Training Instructor, describes how she viewed newly arriving recruits:

> I saw them as people and saw them as human beings. But this was my ball of clay to mold into hopefully, being able to carry on this torch of what I believe the Air Force should be. So I always saw them as my responsibility. [....] And then as they progressed, you would just watch them grow and mature and I would give them more responsibility and they would just run with it. (Blackmarr)

Recall that the Sun's daughter shapes the Two brothers. As Jeff King relates this moment, the Sun's daughter places each on "two things, like rugs" and, using a certain medicine, "pressed [both men] with it all over, molding" and strengthening their bodies (King et al. 45). The correlations are striking. As is the universal necessity that to become part of martial culture, there is a requirement of training and shaping the body, mind, and spirit.

Joining military service today requires a conscious decision to step out of an individualistic culture and transition into a parallel culture. Today's martial cultures face issues traditional cultures such as the *Diné* did not. Initiation within a culture that one is already a part of requires much less transformation than initiation into a different culture. The rites in a collective culture such as the *Diné* reinforce the values, beliefs, and norms of behavior the individual is expected to emulate during his or her formative years. Additionally, each one of the adult members of the culture has passed through the rites, and childhood education is directed toward the successful maturation of the individual as part of the collective enabling a successful passage.

Basic Training must take individuals who grew up with different cultural norms and not only ingrain eight to ten years of understanding and knowledge but also remove habits, prejudices, or other behaviors that work in an individualistic society but are detrimental in a collective culture. Today's US martial cultures make use of concentrated immersion practices to strip various values, beliefs, and norms of behavior that do not align with military cultural requirements.

An advantage typical tribal cultures have is a large amount of ethnic homogeneity. US martial cultures are unique in that the US military organization has one of the most diverse populations with respect to categories such as ethnicity, gender, sexual orientation, and socioeconomic background. Diversity of backgrounds can present another challenge in Basic Training. Those who enter the service with prejudices should find the values, beliefs, and norms of the US martial culture, reinforced through ritual, do not endorse intolerance based on ethnicity, sex, gender, religion, etc. Therefore, Basic Training must first separate the person from individualistic culture, much as a child moves from the world of the womb and emerges into the light of a new world.

This separation phase takes place in a nonordinary physical space, separated from both the culture they are leaving and the one they are going to (Stephenson 29). This specialized spatial limbo immerses the enlistees into the training. Parris Island, Lackland Air Force Base, the Great Lakes, Fort Moore (previously Benning), and Fort Jackson are all liminal space, akin to the dusk of sunset where, like the sun, the enlistees die to their old selves and transform into a new existence.

These arenas of transformation are not part of either the civilian world or the active military force. The individual will enter the culture through these spaces, continually transforming in this liminal space. Each transit—represented by levels of schooling—will vary in length and intensity, but all are meant to redefine the individual by recreating and utilizing the liminal threshold just as the Two brothers are changed by their trials as they journeyed to and tested at the house of the Sun.

Basic Training's primary goals are teaching teamwork, demonstrating that seemingly impossible tasks can be accomplished only through the group and providing a purpose for service that is beyond the self-interests of the individual. More important than being granted a "physical instrument" is to obtain "a psychological commitment and a psychological center" (Campbell and Moyers *Ep. 1*). This commitment and center are the ideals of what each service espouses as core beliefs and values which are to be internalized and then demonstrated in every action and behavior. These behaviors are more akin to the qualities required of a knight of the Arthurian Romance than an obedient serf. The US Air Force, Army, Marine Corps, Navy, and Coast Guard have values and the creeds (mantras) that are required memorization for every recruit. Although these mantras are not always strictly internalized, those who successfully complete Basic Training have demonstrated a basic understanding and a minimal adherence to these values.

A combination of psychological and physiological practices is used to quickly "separate" the recruit who is willingly transitioning from their former culture and life. These practices, as noted earlier make use of neurological plasticity, create the optimal environment and stimuli to enact neurological remapping. Studies have revealed that to "enhance sensory and motor functions" one must place organisms "in complex environments in which there is an opportunity for animals to interact with a changing sensory and social environment and to engage in far more motor activity than regular caging" (Kolb and Gibb). Recent findings in neuroscience demonstrate that establishing a highly stimulated environment in which a mammal is physically engaged results in "enhanced cognitive and motor functions [and that] wide range of sensory and motor experiences can produce long-lasting plastic changes in the brain" (Kolb and Gibb). Just as the Two are subjected to intense testing, many records of older tribal initiation rituals demonstrate the ancients have always made use of the same intense chaos of passing the tests of the threshold guardians and are reenacted by physically moving through the liminal space of Basic Training.

The recruits experience what Van Gennep describes as "rites of separation" (11). These rites include divestment of civilian clothes and investiture of new uniforms and specific hair styles obtained through cutting. "Rites which involve cutting something—especially the first haircut, the shaving of the head, and the rite of putting on clothes for the first time—are generally rites of separation" (Van Gennep 53–54). Further, the trainees are taught how to walk, stand, sit, and eat. They are taught how to speak and the vocabulary and language patterns of their new social group. As Wilhelm van Humboldt describes, language inbounds culture "and *each language draws a magic circle round the people to which it belongs*, a circle from which there is no escape save by stepping out of it into another." (qtd in Cassirer 9 emphasis added)

The recruits are taught proper interaction within the culture and are given a common language, consisting of both verbal and nonverbal cues, in which to communicate. They must learn that no matter their background, socioeconomic status, skin color, religion, or any other divisive identity, they are now to treat each other as teammates with a single language and appearance, separating themselves from what the recruits were before. Together the recruits work to successfully complete the initiation. They have been "un-othered" and made part of the people.

The training very much exemplifies a Durkheimian definition of ritual: "an inherently conservative institution that joins people into a collective and encourages them to look to the past for models and guidance" (Stephenson 38–40): the services' history of heroes, pioneers, great achievements, and blunders. This ancestral grounding is found within the *Enemy Way* ceremony as the *Diné* find it absolutely necessary to recount the history of their people back to the beginning of the cosmos and in US martial culture this is exemplified in the history and symbols of the martial culture. The recruits will be expected to know and internalize the martial culture's cosmological understandings. The recruits are the inheritors of this tradition and are to accept it as their own including the teaching of language, behaviors, and constant enforcement of values and beliefs of the culture.

When the gods (yei) returned at the end of four days, the boys had grown to be the size of ordinary boys of twelve years of age. The gods said to them: "Boys, we have come to have a race with you." So a race was arranged that should go all around a neighboring mountain, and the four started, —two boys and two yei. Before the long race was half done the boys, who ran fast, began to flag, and the gods, who were still fresh, got behind them and scourged the lads with twigs of mountain mahogany. [Talking God] won the race, and the boys came home rubbing their sore backs. When the gods left they promised to return at the end of another period of four days. (Matthews and McNeley 106)

In most cases, failure is a far more effective teacher than success. Exposing that the recruit has inadequate preparation for the test is a benevolent function of the monster, the threshold guardians,[8] the instructors. For the instructor cannot bestow will; self-determination must be conjured from within. The Two are inspired by Little Wind to rise each day, using their failure to fuel their progress, rather than letting it consume them. The purpose of the failures is to inspire renewed efforts to be ready the next time the Two face the gods.

The initiates do not see the hopes the instructors harbor for their success. Yet all initiates will, at some point, find help when they are flagging.

The threshold guardians appear in many guises and in the above passage they come forth to challenge the two boys. As Campbell posits, "The first stage [...] is leaving the realm of light, [...] and moving toward the threshold. And it's at the threshold that the monster of the abyss comes to meet him. [...] the hero is cut to pieces and descends into the abyss in fragments, to be resurrected" (Campbell and Moyers *Ep. 1*). It is vital that a person fail so as to be resurrected into a new form of existence. As Major General Orlando Ward notes in the preface to the US Army's study of the fall of the Philippines to the Japanese in World War II:

The soldier reading these pages would do well to reflect on the wisdom of the statement exhibited in a Japanese shrine: "Woe unto him who has not tasted defeat." Victory too often leads to overconfidence and erases the memory of mistakes. Defeat brings into sharp focus the causes that led to failure and provides a fruitful field of study for those soldiers and laymen who seek in the past lessons for the future. (Morton vii)

Van Gennep refers to these as "rites of transformation" (11). The individual must confront their own inabilities and work to come back to face the challenge again. Failure forces one to look into the abyss of the self and face the need to endure the painful growth associated with getting up each day and continuing to push oneself.

As in past descriptions of tribal initiation rites where the adults come to steal the children away, waving bull-roarers, dressed as deities, and psychologically distressing both the children and parents, the instructors of the US martial culture come forth in full fury. Bull horns replace bull-roarers, the instructors are now deity-like models of the ideal service person, who exemplify perfect uniforms, physical fitness, stamina, attention to detail, and rigid discipline.

Sergeant First Class Pugh was my drill instructor [....] I'll never forget that guy. Just so precise in everything he did. In the language and the way he did drill, in the way that he addressed the soldiers how he took care of you. [....] It was just amazing to me. Sergeant First Class Pugh is who I aspired to be." (Rodriguez)

The instructors are handpicked and trained to take the initiates through their ordeal, as harsh guides and mentors. The Drill Instructors create an initial environment where there is no right answer, and the trainee must find mental purchase within the cacophony of multiple seemingly disembodied voices shouting contradictory messages from different directions. Success is impossible; failure is inevitable. Like Talking God and Water Sprinkler, they imbue both the harsh threshold guardians wielding flaming sword and piercing gaze and, like Little Wind, the inspirationally helpful Athene-as-Mentor or Buddha in the fear-not pose.

The instructors harbor a hidden desire that all the initiates who begin will successfully finish Basic Training but not everyone who begins will successfully complete the move from separation to transition rites. When needed, the instructors can, in quiet moments, function as Little Wind, with a quiet word of encouragement to inspire the recruits to push beyond their self-imposed boundaries and also be pleased by the successes of their charges.

In the second phase of today's training sequence within US martial culture, the recruit moves past fear and preconceptions of what is valuable and what is not as they seek to learn from and become like their instructors. As former Marine Scout Sniper Freddy Osuna, a Pascua Yaqui tribal member and owner of Greenside Tracking notes:

there's a transition that takes place. You don't see it when you're going through it, right. It's just like losing weight over a long period of time, you can't tell. But, for instance, I remember the look in the eyes of my senior drill instructor when I went when I first met him, you know, a month into boot camp, as opposed to the last couple of weeks boot camp. He was standing there one time at the end of the chow line. And as every recruit was going through with their tray, he was just staring at him with a smile. just nodding his head in in, you know, in agreement with what he had done so far, I guess. [....] I could see that he saw something that we didn't see. (Osuna)

As the Basic Training process continues, recruits are taught the minimum of what is needed in martial culture. Mastery is not as important as acceptable functionality. Sustaining rites will be used later to continue the inclusion. Bodies are shaped through constant and demanding exercise. Successful completion of the transformation includes passing physical fitness tests, written and oral exams on history, basic combat skills, even descending into the "cave" of a tear gas chamber to test skills necessary to survive the ordeal of exposure to more lethal substances they may encounter. The trainees go into the wilderness (field exercises) to face a myriad of tests. The obstacle course looms over them as a modern Siegfried's dragon. Each obstacle brings them close to cultural suitability and/or the final "rites of incorporation" (Van Gennep 11).

It is not the obstacle course, gas chamber, or other trials that move out of the way of the recruit; it is the recruit that moves, thinks, and feels their way through each test. Each successive challenge breaks through internal barriers, prejudices, limitations, and expectations. If trainees falter, they are buoyed by the team around them (and surprising to most, their instructors who embody Little Wind) to keep going. As the Two had to race for four days around mountains, the recruits race their old selves, shedding confining limitations. It is very much an internal struggle (triggered by external stimuli) to exceed and master the body and mind. The successful trainee will exit the ordeal with a new physiology and new psychology. As multiple interviewees noted, there is a physical transformation that takes place which unlocks potentialities.

> When I went in to boot camp I was 118 pounds soaking wet. I came out 127 pounds. They had to put me on a meal plan [....] They used to drop me [for pushups] all the time because I'd have a silly grin on my face. [....] I ran my two miles for the final basic APFT [Army Physical Fitness Test] in 10 minutes and 23 seconds. That was the fastest I had ever run it. I came in second place out of the whole barracks. That day, one guy who was a good runner, said "Stick with me and we'll get this done under 10:30." We did it. It was just amazing, an amazing feeling. (Rodriguez)

The permanence of the change at this stage is tenuous. Basic Training recapitulates the act of growing up. All the recruits have really learned is equivalent to childhood understanding of and interaction with the culture. The Two brothers who have succeeded in racing the gods as well as the graduate of Basic Training have gained much, yet the next initiation into full adulthood must be met.

When a recruit has successfully completed Basic Training, there is another liminal space or series of spaces the individual must pass through. To become a full-fledged member of martial culture, they must go through the equivalent of a puberty rite, where they learn their function as an adult. The next phase consists of learning and qualifying in the role the recruit will have within the martial culture. This rite of passage can be either one school or series of schools that the recruit-initiate must successfully pass to enter the culture as full-fledged members of the martial world. These schools are commonly referred to as Advanced Individual Training, A-School, Technical School, and follow-on training that can last from a few months to nearly two years. This stage is yet another complete departure, transformation, and return as the recruits are separated from their original basic training group and placed into smaller social groups categorized by their functional role within the military structure. Again the recruits are led by guides, this time specializing in this subset (band) of the overall tribal culture. Those who graduate past this point are allowed to enter the martial culture as members. This phase recapitulates ages 8–16 years of age in what is typically thought of as a tribal group. These rites of passage are somewhat analogous to initiation rituals, out of which a person will emerge with a new adult body and in a new adult role.

> I had gone in 140 pounds, came out 165 pounds. Just pure muscle [....] Something really happened here. We can't see it. [....] I noticed people treating me differently and it's probably because I carried myself taller. My posture had changed, my nonverbal communication had changed, [...] I carry myself differently. (Osuna)

In the military, once a person completes their courses, their uniform will reflect the adult role in the culture through distinctive insignia or clothing. Similarly, the recruit leaves their basic training instructors to move on to a more advanced school(s) that will teach and test their abilities to take on adult roles inside martial culture.

As demonstrated in the Two's encounter with Sand Dune Boy, a fundamental point that many fail to understand is that while martial culture trains and prepares individuals to use force, violence is not the only way martial cultures engage with an opponent. Force is only used if presence and dialogue do not yield the desired results. The primary goal of standing military forces is actually to deter conflict. That is, they exist to discourage anyone in the use of force against the culture.

Further, many units in the US military are designed specifically to engage through dialogue and aid rather than with force. As examples, the US Army Civil Affairs Branch and the US Air Force Contingency Response Groups are specifically designed to provide humanitarian assistance to peoples who are not US citizens but need aid in dangerous situations. Many modern military operations range across a spectrum of responses; from natural disaster relief (fighting forest fires, typhoon/earthquake recovery, etc.) to providing life-saving medical evacuation on civilian ships hundreds of miles from shore. To sit down and attempt dialogue with an active, potentially hostile adversary is extremely demanding, but all part of martial culture.

Another example is the widespread unrest and violent clashes between law enforcement and the civilian communities experienced in the United States in 2020. Many National Guard units were activated to restore peace. Some people worried about a repeat of the violence experienced at Kent State in 1970. Instead, reports came back that National Guard units brought a sense of calm and attempted connection. In Atlanta, Georgia members of a National Guard unit were captured on video dancing with protesters while in Minnesota a National Guard unit commander came forward and knelt with the protesters, actively engaging them in conversation (Haney; Cox). Engagement means seeking to achieve an agreeable (and survivable) outcome.

The function of US martial culture to preserve peace is not necessarily recognized or even acknowledged by many in the US military. Part of the maturation of the person within the martial culture is to acknowledge that physical force is on the far end of a spectrum of ways to achieve the collective goals of the larger society. Martial culture is not a blunt object with only one use. Military organizations serve their populations in several ways; going to war is only one of them. When it comes to the two brothers engaging with Sand Dune Boy, their first choice of dialogue allows their safe passage.

Spiderwoman allows the Two brothers to make the mistake of not taking the path correctly, therefore allowing them to learn from their mistakes. In training, martial practitioners often need to learn difficult lessons through the pain of failing to heed their instructors. Proper mentors know when to allow failure. The cost seems severe, yet learning this lesson in training spares the recruit fatally learning it later. Training is the place to make mistakes.

Most schools in the military begin slowly, with both the pace and intensity increasing over time to culminate in an ordeal—a capstone event that pushes the students past what they have known and learned, forcing the trainees, who are about to be sanctified as ready to take their place in the culture, up to and perhaps beyond their breaking point. As a musical score may slowly find the crescendo rising to the climax of the composition, certain courses save the most extreme tests for the end. In many cases, this is where the real lessons are learned.

Advanced military training is not safe. "In 2017, nearly four times as many members of the military died in training accidents as were killed in combat. In all, 21 Servicemembers died in combat while 80 died as a result of noncombat training-related accidents" (Chairman's Mark Summary FY19). While some of these can be prevented, martial training is crafted to prepare recruits and servicemembers for the worst that all the deities of war can conjure. Military training is informed by what those who have gone before in martial cultures identify as things that allowed them to confront war. Not a metaphorical war, but a real, monstrous, subliminal, terrific war. And not to just survive in this environment, but thrive as they complete their tasks and return. A common refrain used as a mantra for many schools is "the more one sweats in peace, the less one bleeds in war." Training should be harder than what may occur when a person is deployed forward. There is always an element of danger when pushing the limits. Experienced and knowledgeable instructors, like the *Diné* gods, Old Age, and Spider Woman will, for the benefit of those recruits under their care, act as hard and unyielding threshold guardians.

However, they also simultaneously fill the role of protective guides. By establishing rigorous standards, the instructors attempt to ensure that only those who have the abilities to survive the fire of conflict are allowed pass. There is a kindness and mercy in telling a student they failed. The instructor, acting as a threshold guardian, lets the individual know they are not ready and are quite possibly protecting the individual from physical harm. Additionally, an instructor is exhibiting a wider communal love by ensuring that those who do pass will be able to rely on the person next to them. Likewise, the Sun, holding the knowledge of what the Two may face, begins a series of tests for weaknesses that, should they be allowed to pass with them, will doom both the Two and the entire *Diné*.

In military organizations today, all members are paired during basic training. The recruit will not be more than an arm's length away from the person they are paired with at any time. What happens to one, both will experience. In time the pair of recruits will join with another pair, and these four will create one of the most basic organizational groups in the military, known as a team. The description of the Sun summoning a second pair to join the Two is an almost perfect mythic example of the formation of a team in the US military today. One recruit joins the military and in Basic Training is given a "sibling." The two recruits are trained together and, when it is deemed appropriate, are joined with another pair.

Battle

In the myth, the Two—alighting on Turquoise Mountain—must take a moment, allowing the dark of night to give way to the breaking of dawn and awakening of the earth before moving forward. Stepping onto the soil of a potential battlefield for the first time can be a kind of grounding that, if reflected upon, represents a key spiritual moment. To breathe in the air, feel the wind, let the eyes adjust to the landscape, to assess and bring that place within oneself, is a seminal moment for many the first time they experience it. Some servicemembers are completely unaware the moment exists and fumble past it, only to realize later the poignancy of the stillness that exists as one sets foot on to the land that bears the weight of their duty.

Beyond the act of standing, there is a receptivity a servicemember asks from the land. There is a technique/ritual that is taught to certain members of US martial cultures. When on a mission or patrol, the group arrives to a given location via parachute, helicopter, ground vehicle, or by foot. Wherever the noise and commotion of arrival is experienced, the servicemembers are taught to conduct what is referred to as a SLLS (Stop, Look, Listen, Smell) halt. This means quietly separating from each other, forming a circle facing out in separate directions, and silently going prone (laying down on their bellies). The group now goes quiet for five to ten minutes, allowing their hearts and breathing to slow, stretching out the senses to connect with the world and feeling the natural world to return to its rhythms, thereby allowing the team to become part of those rhythms. When the time is right the team silently looks to each other, quietly rise, and move out as one. In an age of people clamoring to teach mindfulness to martial culture, the culture itself has taught this practice to each other for hundreds of years.

The Two, upon the breaking of the dawn, go through a type of ritual by placing their live-eagle feathers to their hearts, donning their flint armor, receiving the lightning arrows and other weapons, and checking the medicine bundle the Sun gave them. US servicemembers are taught from the beginning how to prepare for each day. Laying out their equipment to ensure that nothing is missing or broken. Servicemembers who go to the field on exercise or have deployed thoroughly understand the activity of putting on their armor and picking up their weapons. This steady and deliberate process happens each time one prepares to move into a place of potential danger. It becomes routine, sometimes done seemingly mindlessly. The servicemember checks their personal first aid kit, ensuring the combat tourniquet device(s) is easily reachable… just in case. The preparation for going forth is an old ritual; how old is left to the imagination. It is something everyone in martial culture personally understands, whether they are an engine mechanic, an air traffic controller, or a Marine infantryman.

It should be recognized that throughout their conception, birth, maturation, training, experiences of pain, and deprivation and mortal danger, the experiences of the Two brothers have not served to make them more prosperous or safer. And nothing has been promised, no earthly rewards nor guaranteed place in the world below. The Two are literally created (as is their mother) solely to meet the needs of the *Diné*. When this ceremony is performed and/or told, do the *Diné* recognize the suffering and deprivation experienced by the Two? The brothers continue on to each challenge, it seems, with a glad heart and an unshakeable conviction (though the story does show moments where they lose hope of surviving their training/tests). The Two do not seem to ask, "What is in it for me?" They simply keep moving forward, albeit with aid from many, but ultimately, the brothers must face these challenges alone.

Like the training and tests the Two endured, military training is designed to increase the servicemember's capabilities and thresholds for physical, mental, and emotional discomfort to allow the individual to work smoothly in the chaos of anything from high-stress tasks to natural disasters to combat. As such, "Military operations […] expose individuals to multiple stressors, including sleep loss, food deprivation, and sustained physical [and cognitive] activity" (Lieberman et al. 929). The training must be of suitable intensity and duration to achieve the intended affects. The person volunteers to place themselves into a world of deprivation and suffering yet somehow seem to crave the martial lifeway. Nothing extrinsic is rewarded other than the approval of the instructors and adoption into a smaller tribal group of those who also have shared in the deprivations. These experiences are more than just achievement, and they exist as breakthroughs of spiritual experience, epiphanies, and peak experiences allowing for a momentary glimpse of the transcendent. As Bill Moyers notes, The Koran speaks, "Do you think that you shall enter the garden of bliss without such trials as come to those who passed before you?" (Campbell and Moyers *Ep. 1*).

The valleys are the moments of despair, like those experienced by the Two when they feared they would die in attempting the trials. The peaks are the gardens of bliss, where the Two are in full power with the other pair as they face their father the Sun, invested with the armor, the uniform of their lineage that is not given, but earned.

The battle with Big Giant is the first and most significant engagement the Two participate in. Little Wind, who had advised them since they were children, beginning with how to win the race with the gods, is still present, whispering in their ears, perhaps even in their own internal voices, how and when to coordinate their actions with both each other and Sun. Each servicemember has experienced the manifestation of Little Wind: a kind word of encouragement when the recruit is wavering during training or a quiet voice inside that creates a still center in the midst of the chaos flying around one in heat-baked afternoon streets of Mogadishu, Somalia, or the freezing jagged peaks of Afghanistan. Little Wind is the voice of the ancestors and the intuition and awareness that arises just as it is needed. Little Wind comes again in moments of need, reminding servicemembers of the ancestral struggles at Bataan, Chosin, Normandy, and countless other battles, inspiring not only hope, but those who must root themselves in the martial ancestral soil represented by the blood, sweat, and lives sewed by those who came before.

This is the hope of every martial member in terms of fighting an opponent. A battle can be won, but to secure peace, steps must be taken to ensure the opponent will not regain the strength to fight again later. The hope is that, if there is to be a conflict, the fight will only need to happen once. Containing the blood and destroying Big Giant's cap work to keep the *Diné* safe from having to continue a cycle of endemic warfare against Big Giant. This is not an easy task in today's US martial culture. The 2011 withdrawal from Iraq precipitated events culminating in US forces returning to Iraq and engaging in Syria. How to set the conditions so that a peace may be secure is something that has challenged martial cultures throughout time.

Embodied Lifeway

Singing while walking on foot or marching is a root behavior in martial culture. William H. McNeill, in describing his personal experiences of drilling as a US Army draftee in 1941 writes, "Words are inadequate to describe the emotion aroused by the prolonged movement in unison [...] and every group that keeps together in time, moving big muscles together and chanting, singing, or shouting rhythmically" (2) finds a state of heightened consciousness and a losing of self to the collective. Singing or chanting in time to movement can create a heightened state of being in which the participants find endorphin-releasing exaltation which relieves the feelings of pain, hunger, fatigue, and even the sense of time. As the gods gave the Two songs to give them strength, so to do Training Instructors give songs to recruits in US martial culture today.

Endemic Warfare

For the US servicemember, intermittent yet endemic warfare has been a constant, especially since the end of the Cold War: including Operations Desert Shield/ Storm, Northern Watch, Southern Watch, the interventions in Somalia, Bosnia, and Kosovo to name the larger demands on the US military. That operations tempo increased dramatically in September 2001, when the same servicemember would have brief moments at home while rotating between Bosnia, Kosovo, Afghanistan, Iraq, among many other regions around the world. "Between 9/11 and September 2015, 2.77 million servicemembers served on more than 5.4 million deployments" (Wenger et al. 1) ranging from 1 to 18 months in length with a typical US Army length of 12 months. Further, approximately 610,000 servicemembers served three or more deployments (Wenger et al. 8). As Marine Sgt. James Dunham, an Apache puts it "Now I get restless if I am home too long, knowing that there is still the fire within me to fight and my brother Marines are still fighting […] I have deployed to Iraq in 2003, 2005, and now 2007" (Viola 161–62) What is unique about these servicemembers is that each volunteered to continue to stay in the martial culture; mirroring the experiences of the Two.

In the myth, the Two find themselves in the repetitive lifeway of endemic battles. The story of the Two was born out of a similar need to understand and psychologically center those who participated in this demanding lifeway of endemic conflict (Tedlock 263–64; Brunt). The *Enemy Way* describes the future for all young adults, irrespective of gender, who will experience a continuity of conflict and return. While the *Enemy Way* ceremony is about two brothers, and about all Navajo warriors today who serve, its import and universal song is vitally relevant to all those who serve within US martial culture.

The Two apply the hard-won knowledge and experience to lead when needed and that accumulated knowledge and wisdom is both recognized and accepted as valuable to the *Diné*. The *Diné* have a martial culture and therefore understand what the Two can contribute. The *Diné* attitude is quite different from what veterans outside of the American Indigenous Nations face today when one moves completely out of martial culture and anything learned within the military organization is considered useless and even detrimental. The veteran and their family must start over, going through academic institutions and gaining degrees recognized as valuable in civilian culture. Yet just as the Two will always be warriors for the *Diné*, the veteran will always be of the martial culture. The hands, legs, mind, and heart do not forget even as they find they have nothing further to give because the civilian culture wants nothing from them beyond acquiescence.

Cleansing and Living Forward

"When someone has gone into combat, they need to be spiritually and emotionally cleansed" (qtd in Simkins and Barrett). The Enemy Way Ceremony depicts the wear living a martial lifeway can have on a person. In the myth, the balm and renewal is the telling of the Two brother's story… beginning with the origin of their people and ending at the moment the Two are currently in. It is their story, told to them by their spiritual elders of their people. And it is recounted four times. And only in this way, can the Two speak "of living in the future, and of the making of the future people" (King et al. 52).

Unfortunately, true rituals that effectively bring servicemembers back into alignment with their martial cultural heritage, thereby centering their inner selves, are lacking. However, what is much more detrimental is the widespread lack of authentic rituals for those servicemembers and families who leave the immersive military organization. Most simply separate after their contract is complete having no official ritual associated with departure. They are secularly separated from the military organization in perfect administrative fashion, as if simply finalizing paperwork on purchasing a home. However, the secular separation is false. The person will forever be part of the martial culture from which the secular government and civilian culture seeks to remove them. Yet, the paradox is that the separation is also a very real excommunication. Whether by choice or by need, the member of the martial culture is ejected into another world. At the end of the Vietnam War many of those who separated from the military organization failed in procuring work of any sort due to lack of skills and heavy discrimination (Shay 51).

The servicemember-turned-veteran may adapt; the current word utilized for the adaptation is *transition*. But to do so, many believe adaptation requires a suppression of a great deal of the veteran's martial cultural values, beliefs, and norms of behavior. A formal recognition of martial culture will acknowledge the lifelong membership to service, normalize the reality of the holistic nature of martial culture which reaches into all aspects of an individual, and allow for the space to create the conditions necessary to integrate those positive aspects of martial culture into the continued growth of the individual.

The *Enemy Way* ceremony is not just a metaphor but a living mythos reenacted every time a person becomes a member of martial culture. The ritual imparts the meaning of service and its connection with history, the purpose of initiation and acts as a guide to model a servicemember's actions. The realization that back-to-back deployments are not new, but a far older martial-cultural norm, would benefit many who are not aware of the reality of endemic conflict. The *Diné Enemy Way* ceremony is a vital link to the US collective martial-cultural past and may guide servicemembers and veterans how to move forward together.

Notes

1 *Diné*, while generally interpreted as "the People," is really a combination of two words. *Di* "means up where there is no surface" and *né* "means down to the surface of Mother Earth so we refer to ourselves as people or beings that came from a place […] that had no surface to the earth" ("Navajo Teachings: Fake History vs. Real History.").

2 Grace McNeley praises Matthews' works, considering them vital resources to not only the academic community, but also for "Navajo people in our ongoing efforts to preserve and strengthen our cultural traditions" (Matthews and McNeley xiv–xv).
3 This small detail is practiced within the US military. The lower-ranking individual walks to the left of the higher-ranking individual.
4 The Sun is referred to as both the boys' father regardless of the fact that the Sun is the older boy's father and the grandfather of the younger, who was born of the Dripping Pool.
5 Campbell argues "Now you might say this is cheating, but it isn't cheating. If they weren't worthy of this, they would not receive" the aid (Campbell *Transformations of Myth* 42).
6 Four, as King relates, is the most powerful number to the *Diné*, as is clearly depicted in the cycles of four illustrated throughout the myth as well as the Four Holy Mountains that circumscribe the *Diné's* home (King, et al. 46).
7 The importance of the number four in Diné is so important that the two must be four, even though Reared Underground and Changing Grandchild never appear again in this story.
8 Threshold guardians serve to keep individuals who do not meet the holistic requirements to pass from one dimension to another. Often, they are represented at temples as fierce monster figures. In mythology, monsters who guard linear obstacles such as river crossings appear as threshold guardians. They are meant to be overcome, but can only be pacified by one who has the unique qualities necessary to pass. They are fierce judges who exist to test those who present themselves for assessment.

References

Blackmarr, Jennifer. Personal Interview. 9 July 2023.

Brunt, Charles D. "Two Navajo Women May Have Been America's First GI Janes." *Albuquerque Journal*, www.abqjournal.com/887174/were-navajo-women-first-gi-janes.html, accessed 12 Aug. 2020.

Campbell, Joseph. *Transformations of Myth through Time*. Harper and Row, 1990.

Campbell, Joseph, and Bill Moyers. "Ep. 1: Joseph Campbell and the Power of Myth 'The Hero's Adventure.'" 27 Aug. 2018, https://billmoyers.com/content/ep-1-joseph-campbell-and-the-power-of-myth-the-hero%E2%80%99s-adventure-audio/

———. "Ep. 6: Joseph Campbell and the Power of Myth "Masks of Eternity."" *BillMoyers.com*, 19 Oct. 2020, https://billmoyers.com/content/ep-6-joseph-campbell-and-the-power-of-myth-masks-of-eternity-audio/

Cassirer, Ernst. *Language and Myth*, translated by Susanne K. Langer, Dover Publications Inc., 1953.

Chairman's Mark Summary FY19 NDAA V1. Reform and Rebuild: National Defense Authorization Act for FY2019, https://docs.house.gov/meetings/AS/AS00/20180509/108275/HMKP-115-AS00-20180509-SD001.pdf

Cox, Matthew. "Guard Won't Punish Soldiers Who Took a Knee at Protests." *Military.Com*, 8 June 2020, www.military.com/daily-news/2020/06/08/guard-wont-punish-soldiers-who-took-knee-protests.html

Haney, Addie. "Watch: National Guard Troops Dance in the Streets with Protesters." *11Alive.Com*, 5 June 2020, www.11alive.com/article/news/local/protests/atlanta-protests-national-guard-dances-with-crowd/85-47d916ed-dc79-4457-a646-93aa5c74b5be

Kearney, Richard. *Strangers, Gods and Monsters: Ideas of Otherness*. Routledge, 2003.

King, Jeff, et al. *Where the Two Came to Their Father: A Navaho War Ceremonial*. Princeton UP, 1991.

Klah, Hasteen, and Mary C. Wheelwright. *Navajo Creation Myth: The Story of the Emergence*. AMS Press, 1980.

Kolb, Bryan, and Robbin Gibb. "Brain Plasticity and Behaviour in the Developing Brain." *Journal of the Canadian Academy of Child and Adolescent Psychiatry = Journal de l'Academie canadienne de psychiatrie de l'enfant et de l'adolescent*, vol. 20, no. 4, 2011, pp. 265–76.

Lieberman, Harris R., et al. "Cognition during Sustained Operations: Comparison of a Laboratory Simulation to Field Studies." *Aviation, Space, and Environmental Medicine*, vol. 77, no. 9, 2006, pp. 929–35.

Maryboy, Nancy C., and David Begay. "The Navajos of Utah." *History Of Utah's American Indians*, edited by Forrest S. Cuch, et al., UP of Colorado, 2000, pp. 265–313. *JSTOR*, www.jstor.org/stable/j.ctt46nwms.10, accessed 18 Mar. 2021.

Matthews, Washington and Grace A. McNeley. *Navaho Legends*. U of Utah P, 1994.

Morton, Louis. *The Fall of the Philippines*. Center of Military History, United States Army, 1993.

"Navajo Teachings: Fake History vs. Real History." *YouTube*, uploaded by Navajo Traditional Teachings, 12 July 2021, www.youtube.com/watch?v=MXpJd-V2n4U

Osuna, Freddy. Personal Interview. 14 June 2023.

Rodriguez, Riccardo. Personal Interview. 16 June 2023.

Sapolsky, Robert M. *Behave: The Biology of Humans at Our Best and Worst*. Penguin Books, 2018.

Schilling, Vincent. "By the Numbers: A Look at Native Enlistment during the Major Wars." *Indian Country Today*, 6 Feb. 2014, www.indiancountrytoday.com/archive/by-the-numbers-a-look-at-native-enlistment-during-the-major-wars?redir=1

Shay, Jonathan. *Achilles in Vietnam: Combat Trauma and the Undoing of Character*. Scribner, 2003.

Simkins, J. D., and Claire Barrett. "A 'Warrior Tradition': Why Native Americans Continue Fighting for the Same Government That Tried to Wipe Them Out." *Military Times*, 15 Nov. 2019, www.militarytimes.com/off-duty/military-culture/2019/11/15/a-warrior-tradition-why-native-americans-continue-fighting-for-the-same-government-that-tried-to-wipe-them-out/

Stephenson, Barry. *Ritual: A Very Short Introduction*. Oxford UP, 2015.

Tedlock, Barbara. *The Woman in the Shamans Body: Reclaiming the Feminine in Religion and Medicine*. Bantam Books, 2006.

Van Gennep, Arnold. *The Rites of Passage*. University of Chicago, 1960.

Viola, Herman Joseph. *Warriors in Uniform: The Legacy of American Indian Heroism*. National Geographic, 2008.

Wenger, Jennie W., et al. "Examination of Recent Deployment Experience across the Services and Components." *RAND Corporation*, 2018, https://www.rand.org/pubs/research_reports/RR1928.html

Woodard, Colin. *American Nations: A History of the Eleven Rival Regional Cultures of North America*. Penguin, 2011.

4   Hindu

The *Mahabharata* is a Hindu epic that describes the entire life cycle of warriors (*kshatriya*) with focus on a family torn apart by a cosmic war between gods and demons. The *Mahabharata*—and the *Bhagavad Gita* contained within—are spiritual, psychological, and philosophical guides to millions who seek "to live a human lifetime under any circumstances" (Campbell and Moyers *Ep. 2*). This mythology can also be approached as one of the spiritual inheritances of martial cultures today. This chapter is a brief exploration of just a few of the touchpoints that may resonate with servicemembers, veterans, and their families of US martial culture.

The numerous diversities of interpretations for this seminal Hindu text exist which complicates efforts to find a translation free of biases by the interpreter or reteller. For example, Mahatma Gandhi took the view that the text is allegorical (Gandhi and Desai 9–10) and should be approached as a guide to right action in life generally, not as a war epic. George Feuerstein, while admitting the text could be read literally, insists one should ignore the "militaristic" aspects of the *Bhagavad Gita* and fully embrace a metaphorical reading (Feuerstein and Feuerstein ix). While these are valid lenses, the *Mahabharata* is also an applicable mythos of the warrior life cycle and has many lessons from ancient wisdom about how to live and formulate an adherence to martial culture.

The *Mahabharata* also serves a sociological function by upholding a certain cultural view of how Indian society should function, which at many points creates ethical and moral cognitive dissonance for those readers who espouse differing values, beliefs, and norms. The purpose of this chapter is to look at the Hindu myth through the lens of martial cultural theory and see what touchpoints emerge that may speak to US martial culture today.

The mythological symbolism of the *Mahabharata* could describe any number of wars; however, most scholars believe the text refers to a specific historical battle that took place at Kurukshetra in the modern state of Haryana in northern India near New Delhi. While archeological data suggest the date of the battle itself is likely c. 1500 BCE, there is a tradition that claims the war ended in 3102

DOI: 10.4324/9781032613222-4

Feb 18 BCE (Feuerstein and Feuerstein 18–19; Venugopal). Archeologists have attempted to determine the historical veracity of the text and have discovered numerous items, including war chariots and other detritus consistent with wars of the time (Venugopal). The descriptions of the myriad of experiences attributed to the warriors of the *Mahabharata* suggest that the contributing author—be it the great Vyasa to whom it is credited or a series of other authors who added to the mythology over time—understood or perhaps lived as *kshatriya*. Regardless of the source or sources that contributed to the *Mahabharata*, the myth has a great deal to teach the US servicemembers.

The prevailing theme explored in this chapter is the crisis of duty and service weighed against personal comfort, self-interest, or pleasure. In exploring the *Mahabharata*, nearly every page speaks to the life cycle of the warrior including training, familial sacrifice, the loss of martial family, and seeking the guidance of elders. The chapter explores the *Mahabharata* as a frame tale—actually many tales embedded within a single tale—told as a history of the great schism between the Pandavas and the Kauravas. It demonstrates correlations in the US martial culture. The Mahabharata largely focuses on themes regarding relationships between teacher and student, maintaining personal duty by adherence to accepted cornerstone values, moral injury, and living forward. The Hindu epic is the hymn of martial ancestors singing forward in time to guide those who fulfill warrior roles today.

When the oral tradition of the Mahabharata began, Hindu society consisted of four distinct castes,[1] each of which was responsible for a different set of social duties collectively understood as *dharma*.[2] The *kshatriya* were principally responsible for leading and providing the warriors for the society. Devotion to *dharma* was believed to result in accumulation of positive karma and a favorable rebirth to a higher station, a step closer to *moksha*, or freedom and departure from the cosmic cycle of *samsara*, the wheel of death and rebirth. The great poet Vyasa, endowed with powers unique to one chosen as the emissary of the gods, speaks the *Mahabharata*. The result is a tale of martial culture and a civil war that contains an entirety of human experiences which can guide US servicemembers and veterans.

The Mahabharata

A brahmin warrior wages battle, killing all *kshatriya* men. To continue the line of *kshatriya*, *kshatriya* (warrior) women conceive children with *brahmin* (priests) men. The creation of warrior-priests heralds a golden age of reverence for *dharma*. People live in peaceful cooperation, without the negative emotions that beget disunity. All living things flourish as nature intends.

The golden age ends when demons reincarnate on earth as corrupt *kshatriyas* who begin an era of lawlessness, attacking the weak, and destroying the world itself. The Earth beseeches Brahma, Lord of Creation, "O Lord Brahma, I am overwhelmed by so much wickedness. I shall be destroyed!" (Satyamurti 9). Brahma gathers all the gods and instructs them to use "*a portion of yourselves to endow a human being with god-like power. Employ your attributes as you see fit. Pitch your strength against the demonic forces which threaten to engulf the entire earth*" (Satyamurti 9). The Pandavas, who are the children of these gods, are charged to defend the earth itself.

The granduncle of the Pandavas is Bhishma.[3] Throughout his life, he exemplifies fulfilling his *dharma* or duty and is beloved by all. As a young man Bhishma makes a solemn vow to honor his father: "here and now, in the name of all that I hold sacred, in the name of my guru, of my mother, and of dharma, I vow to live a life of celibacy. I shall never marry" (Satyamurti 17–18). Bhishma devotes his life to advising, teaching, and servitude. He surrenders ambitions of heredity and children. In return, Bhishma's father grants him the ability to choose when he will die. Bhishma, in his oath to his father, follows the ideal of *kshatriya dharma*, which transcends individual loyalty or personal relationships. *Kshatriya dharma* is an immutable set of universal principles derived from the harmonies of cosmic balance, not to be dismissed due to inconvenience or hardship.

Bhishma becomes an uncle to Pandu and Dhritarashtra. Pandu fathers the Pandavas and Dhritarashtra fathers the Kauravas. Pandu marries two wives, Kunti and Madri, while Dhritarashtra, blind from birth, marries Gandhari. Through a series of incidents, Pandu the eldest, retreats into the forest, abdicating the throne to Dhritarashtra. Pandu's two wives, Kunti and Madri, volunteer to accompany their husband into the forest. Meanwhile Gandhari, on her wedding day, binds a cloth over her eyes, choosing to join her husband Dhritarashtra in his sightless condition.

At the request of Pandu, Kunti and Madri pray to be given children by five different gods; Kunti is visited by three gods: Dharma, Vayu, and Indra, and Madri is visited by the twin gods known collectively as the Asvins. The result of these unions are the five Pandava brothers: Yudhishthira, Bhima, Arjuna, Nakula, and Sahadeva. As children of the gods, they embody five different attributes, each of which reflects various aspects of the whole *kshatriya* or warrior self.

Yudhishthira, the son of Dharma, is a reflective, empathic, and feeling leader. Aside from a weakness for gambling, Yudhishthira demonstrates the virtues of leadership and, while open to guidance and advice, wields authority without shirking ultimate responsibility for his decisions. Yudhishthira is the recognized heir to the throne of the kingdom after his uncle, Dhritarashtra, dies.

Bhima, the son of Vayu,[4] exhibits an untamable supernatural wildness and strength, inherited from his father, who is described within the Vedas as a god of strength. Bhima is "tigerlike, enemy-taming [...] of terrible strength" (van Buitenen 299), reflecting the courage, tenacity, and force of nature a warrior has within. He is energetic and full of ferocity, with a keenness to close with an adversary.

Arjuna's father is Indra.[5] Arjuna is the "bull-like [...] Terrifier" (van Buitenen 273), who is somewhat distant, always seeking perfection in his warrior skills and questing for higher knowledge of the arts of combat. As Indra is the god of battle and vanquishes darkness, it is only Arjuna who can prevail against the foes arrayed against the Pandavas. The Asvins are the fathers of the twin boys Nakula and Sahadeva. These brothers are bequeathed "with beauty, courage, and virtue beyond all other men" (van Buitenen 259). Nakula and Sahadeva are also endowed with the powers of a healer through their fathers.

The five Pandavas are raised in the wilderness until Pandu dies. During the ritual burning of his body, his wife Madri joins him on his funeral pyre. Kunti is left to mother her three sons as well as Madri's twins. Thus, Kunti becomes mother to all five Pandavas.

Meanwhile, Gandari, wife to the blind king Dhritarashtra, grants the holy Vyasa shelter and care and in return is given 100 sons. The first-born, Duryodhana, comes into the world at the same hour as Bhima, Kunti's second son. At his birth, Dhritarashtra's brother Vidura proclaims "Oh, my brother, this birth portends the ruin of your line. Your first-born son is destined to destroy all that we've held sacred through the ages" (Satyamurti 34). Dhritarashtra, for the love of his son, allows him to live, despite the threat he represents to the peace of the world. Duryodhana and his 99 brothers are collectively known as the Kauravas.

The final incarnation (child) to be born on earth, Krishna, child of Vishnu, is born on earth and joins the Pandavas and Kauravas. Unlike the others who are unknowingly incarnated deities of their heavenly fathers, Krishna is completely aware he is the incarnated Vishnu. Vishnu's traits help Krishna become counselor and companion to Arjuna.

After Pandu's and Madri's deaths, Kunti and her five sons rejoin their family in Hastinapura, the capital of the Bharatas, placing themselves under the care of Bhishma. Here, the seeds of jealousy are sown between the Pandavas and Kauravas. The Pandava Bhima's innate character endears him to nearly everyone. However, Duryodhana views the Pandavas as obstacles who stand "between him and life's advantages—power, in particular" (Satyamurti 39). Bhima is considered the primary undefeatable obstacle and Duryodhana plots to remove Bhima and the Pandavas from the line of succession, clearing the way for Duryodhana to become king.

Bhishma exemplifies the idealized pursuit of excellence by his care and protection of all of his grandnephews. As he nurtures the young Pandavas and Kauravas, who all view him as a grandfather, he arranges for their education is focused on training in martial culture and "the arts of warfare" (Satyamurti 44). To this end, Bhishma gives the children two teachers, Kripa and Drona, both of whom are warrior *brahmins*. Kripa establishes the foundational knowledge, then Drona provides training "more advanced in all the branches of the arts of war" (Satyamurti 45).

Drona "had acquired rare and powerful *astras* [supernatural weapons]" (Satyamurti 45). He understands that to become a master in wielding bow or sword requires much more than physical adroitness, great strength, or even perseverance. "Qualities of heart were needed, stillness of mind and body, complete focus"

(Satyamurti 45–46). Further, the appropriate use of *astras* is "dependent on the depth of spiritual maturity attained by the [person] who would summon them" (Satyamurti 51). The Pandavas and Kauravas learn from their teachers that austerity is a necessary experience for warriors before they are allowed to move forward to gain greater knowledge. Those who evidence humility, patience, and the ability to accept responsibility may attain greater power.

The roles of Bhishma, Kripa, and Drona demonstrate a positive teacher-student relationship. Kripa and Drona, as *brahmin* warriors, provide a balancing aspect to the teaching of the Pandavas. Kripa provides the basis upon which Drona builds and instills the highest aspect of action which is the quality of heart and the stillness at the center of mind, body, and soul all aligned into a balanced calm that evidences the harmonious union of both qualities.

Arjuna, the great *kshatriya*, gives himself to the guidance of his *Brahmin* teachers and, in return, taps into the place of flawless action:

> Suddenly, he rose— and running out into the moonless night he flexed his bow, nocked an arrow, let fly, although the target was invisible; [...] he found each arrow clinched into the place he had intended. (Satyamurti 52–53)

Arjuna exemplifies an alignment of physical, mental, and spiritual clarity of self that allows for nonattached action that evidences an enlightened warrior. Arjuna "had a glimpse of how one may become a channel for the world's natural forces to play themselves out. How, without striving, without attachment to the end result, abandoning desire and memory, an arrow can be loosed, and find its home" (Satyamurti 53). Arjuna passes through a threshold upon experiencing perfect action and is granted access to higher teachings. "Not a thought had ruffled Arjuna's mind. He had simply acted" (Satyamurti 54). For this, Drona then grants Arjuna the Brahma Head, "a weapon so deadly it could not be used against mere mortals without burning up the whole world" (Satyamurti 54), a weapon that is only to be used in supernatural combat.

Unbeknownst to the Pandavas, they have an older brother who suffers the life of the outcast warrior. Kunti, before she was wed to Pandu, secretly conceived a son with the sun god Surya. Surya declared the child, Karna would have golden armor and earrings that protect him from all danger. Once born, Kunti placed the babe in a basket on the sacred river Ganga in the middle of the night. Karna is found by a chariot driver, Adhiratha, and his wife Radha. The couple took him in as their own. Karna loved and revered his parents and they cherished him. Yet rather "than being a driver like his father, his natural talents and his inclinations tended toward a hero's martial calling" (Satyamurti 65). Since Karna was born of Kunti and Surya, he is a *kshatriya*. However, ignorant of his beginnings, society casts him in relation to his adopted parents' status.

Adhiratha takes Karna "to train at Drona's weapons school" (65). There Karna faces ridicule and humiliation for his station as a chariot driver's son, despite the fact that his skills are equal to Arjuna's. When Karna asks for the *Brahma* weapon, Drona responds that the weapon is only for "a brahmin of stringent vows, or a kshatriya who has undertaken great austerities" (Satyamurti 56). Karna, by virtue of merit, meets all of the standards to be granted knowledge of the *Brahma* weapon, yet he is denied. Karna leaves Drona's school and completes his education from another teacher and continues to suffer from prejudice based upon the caste of his adoptive parents.

Karna eventually returns and outperforms Arjuna at a tournament that is meant to display Drona's *kshatriya* students. Karna's performance makes Arjuna's anger turn to rage and, loathing that this "son of a wagoner" dares to outdo him, a prince and kshatriya, while Bhima taunts Karna. Duryodhana comes to Karna's defense: "[Bhima], your rudeness and crass ignorance are hardly worthy of the kshatriya you claim to be [.... Karna] has proved himself equal to any of us" (Satyamurti 79). Duryodhana embraces Karna and grants him the kingship of a vassal state as a reward for Karna's skill and courage. The tournament signals the end of the Pandava's and Kaurava's training under Drona.

The Pandavas and Kauravas go forth into the world as *kshatriyas*. In their first battle under Drona's command, the inner natures of the Pandavas and the Kauravas begin to surface in how they comport themselves in combat. The Kauravas, under Duryodhana, are eager to storm the city and are "consumed by feverish excitement" (Satyamurti 82) while the Pandavas, "calm and more thoughtful" (Satyamurti 82), watch as the Kauravas attack the city. Duryodhana seeks to breach the city using a frontal attack and superior numbers. For the Pandavas, "self-restraint was their first victory" (Satyamurti 82). The Pandavas know their adversary and gain knowledge by watching their cousins' assault. The Kauravas attack fails and they "learned that a thirst for victory was not enough" (Satyamurti 82). Four of the five Pandavas then conduct a raid focused on capturing the king of the city. The Pandavas show themselves to be less impetuous and more studied in their approach to warfare. After this first battle, the Pandavas are bolstered in their confidence and continue in their *kshatriya* roles serving the kingdom.

The popularity of Yudhishthira with the people, partly due to their successes in expanding the kingdom, gives the blind king Dhritarashtra concern that they covet his position. Further, Yudhishthira is the rightful heir once Dhritarashtra dies; therefore, Dhritarashtra's son Duryodhana will not inherit the kingdom. This fear and jealousy feeds decisions to subvert or kill the Pandavas. The Pandavas continue to meet each attack with tolerance and forgiveness. The Pandavas seek to find ways of coexisting with the Kauravas, even in contradiction to Krishna's guidance, who alone is aware of the cosmic necessity of the battle to save earth.

These conditions continue to spiral until Duryodhana arranges a tragic game of dice in which Yudhishthira loses, and the wife of the Pandavas, Draupadi, is shamed by the Kauravas. The game ends with the Pandavas exiled. "For twelve years [the Pandavas] will be exiled in the forest; the thirteenth year must be spent in public, incognito. If [they are] recognized, then another thirteen years of exile must begin" (Satyamurti 203). Success will result in the Pandavas regaining their half of the kingdom. Yet Duryodhana's plan is to use those years "to assemble a huge and loyal army, and to garner powerful allies" (Satyamurti 204).

The Pandavas retreat into the forest conducting pilgrimages while also acting as protectors to those who show them kindness. During exile, the Pandavas are aware that Duryodhana intends to retain the entire kingdom regardless of whether or not the Pandavas successfully meet the conditions of the wager. Further, they are aware that "Bhishma, Drona and Kripa, although they love us equally [will use] their skill and their celestial weapons" (Satyamurti 232) in Dhritarashtra's service.

Yudhishthira understands that while conflict should always be avoided, it is dangerous and costly to foresee a potential conflict and do nothing. He charges Arjuna to quest for celestial weapons and knowledge in preparation for war. Preparing for conflict does not mean committing to conflict, and the hope that those preparations will never be necessary should not be abandoned.

Only Arjuna, of all the Pandavas, is invited to pursue the higher knowledge because he is the only one to reach a state of effortless action. Even so, Arjuna is not simply allowed to move into the liminal plane of celestial training. He must first be tested. Once he reaches the mountain Indrakila he is stopped by Indra, disguised as an old ascetic, who says to Arjuna, "Why do you come armed to this holy place? There is no conflict here, no enemies— drop your bow, your arrow-brimming quivers. In this land, you will find serenity; this is where your quest ends" (Satyamurti 235–36). The first test is that of renunciation of the quest, the abandonment of family and duty, and the seeking of serenity and a peaceful individual life. The ascetic asks Arjuna to cease being a *kshatriya*, because that life is hard and filled with sacrifice and pain. In giving a choice between a life of serene existence or acting in the world of hardship, the ascetic offers Arjuna a way out.

Arjuna understands the coming struggles will be much greater than what he has encountered previously. Arjuna bows to the ascetic, insisting he needs his weapons for what lies ahead. Indra reveals himself and offers "heavenly joys" (Satyamurti 236) to his son instead of his quest. Arjuna responds with resolution, "no conceivable delight, no sovereignty, no worlds, no happiness, could deflect me from my chosen path" (Satyamurti 236). After successfully passing several tests, Arjuna is greeted by many other gods who "each gave weapons to the Pandava and foretold victory in the coming war" (Satyamurti 239). Arjuna is the guest of his spiritual father, Indra, for five years, in which he "learned to master many divine weapons, studying their diverse applications and how to call them back" (Satyamurti 240) as well as several other spiritual practices.

Meanwhile, the other Pandavas complete a pilgrimage, "touring the fords on the sacred rivers" (Satyamurti 248) where they "paid reverence, gave gifts to priests, immersed themselves and performed penances" (Satyamurti 252).

Before their final year to be spent in disguise, Yudhishthira tours the forests and streams in the wilderness mountains. Upon preparing to leave, he silently promises to return one day.

At the end of 13 years, the Kauravas find the Pandavas have successfully passed their trials and request the return of their half of the kingdom as agreed. Dhritarashtra, under the influence of his son Duryodhana, refuses to honor the original agreement. Therefore, both Yudhishthira, as leader of the Pandavas, and Duryodhana, as leader of the Kauravas, send out envoys to various kingdoms to raise armies. Still Yudhishthira continues to look for a peaceful solution.

An envoy from Dhritarashtra comes to the Pandavas, emphasizing that the Kauravas have 11 armies against the Pandava's 7. Additionally, the envoy raises the specter of kin killing, and how—regardless of who wins or loses—that act would stain any victory immeasurably. Krishna argues against this, stating, "If someone seizes the land of another out of avarice, then a king's duty is to go to war to set things right. Action is the duty of a kshatriya. […] Mere inertia in the face of such flagrant wrongdoing is not virtue. […] Kshatriya dharma is to protect what's right" (Satyamurti 342). Yudhishthira pleads with Dhritarashtra: "we can live in peace" (Satyamurti 343). To attempt to avert war, Krishna goes to Karna.

Krishna encourages Karna, as the actual first-born son of Kunti and Surya, the sun God, to embrace his birthright as a true *kshatriya* and the first son of the Pandavas. If Karna agrees, as closest friend to Duryodhana he can halt the coming war as well as lay claim as the next in line to the throne, uniting both Pandavas and Kauravas. Karna refuses, having already given his word to Duryodhana. As Yudhishthira's efforts demonstrate, all avenues of peaceful resolution are sought. Unfortunately, they are also expended. The armies gather on the plain of Kurukshetra.

Bhishma, Drona, and Kripa understand that they fight for a cause that is born of envy and thus unjust. Yet they are bound by *dharma* to the blind king Dhritarashtra who feels powerless. His son Duryodhana cannot be dissuaded from creating the war. Bhishma, Kripa, and Drona fight on the side that is populated by the reincarnations of demons in the 100 sons of Dhritarashtra.

On the day of the first battle, each side brings their armies and arrays them on the field. Arjuna asks Krishna to drive his chariot into the center so he may survey both of the massive formations. At the sight of those he knows and loves on both sides, aware of exactly what this war will cause and unable to ignore the cognitive dissonance, Arjuna sinks into an abyss of inaction. "Beholding these kinsmen, O Krishna, assembled together and eager for the fight, […] I do not desire victory, [….] I wish not to slay these though they slay me, […] even for the sake of the sovereignty of the three worlds, what then for the sake of (this) earth?" (Ganguli and Vyāsa *Volume V, Bhishma Parva, Section XXVI* 52). Arjuna is fighting on the side of *dharma* to correct a subversion of the cosmic order; this is a just war from the cosmic standpoint. And Arjuna—while he is questioning this battle—is by no means a reluctant warrior. Yet as Arjuna surveys the armies, he observes those he loves on both sides. He is torn and cannot understand how Bhishma can sit so calmly, as if not comprehending that he will shortly be engaged in deadly combat with Arjuna and the other Pandavas. Arjuna views killing kin as far worse than killing other enemies. "Whether this war brings victory or defeat there will be no occasion for rejoicing" (Satyamurti 402). For Bhishma and Krishna, who understand the subtler ways of *dharma*, kin killing is permissible. Arjuna's concern is not his own death, but that he will have to cause their deaths.

Krishna, however, exposes Arjuna to the world as it really is: an illusion of separateness that one should not cling to, but must act in.

> "Friend, this is unworthy of you. [...] Get to your feet, scourge of your enemies!" [Arjuna,] your doubts sound honorable but they spring from deep misunderstanding. You speak as if this life were all there is. But it is just one brief embodiment of the indestructible, eternal soul. Bodies are born, they flourish, age, and die. [...] Wise people know this, and do not lament. (Satyamurti 400–402)

Krishna then reveals the true nature of the universe, the secrets of *dharma* and karma, and a great deal more. Through this tremendous revelation, and the paradigm shifts that are made, Arjuna finds the strength to stand, pick up his bow, remount the chariot, and take his place in the line of battle.

Over the following 18 days, the Kauravas and Pandavas fight desperately. However, it is the struggle over *dharma* and acting within the values of a true *kshatriya* that will last long after the final arrows fly. Before the battle, Yudhishthira, as head of the Pandava army, approaches Bhishma and Drona, asking for their permission to wage war against them, and further, how to subdue them. Understanding their *dharma*, they give their blessings. These three, Bhishma, Drona, and Karna, are three of the four leaders of the Kauravas forces. Duryodhana is the fourth. By the nineteenth day, Bhishma will be mortally wounded and the other three die during the battle.

The Pandavas are usually framed to be "good" and the Kauravas as "bad" in retellings of the Mahabharata. And yet there are several instances where the Kauravas demonstrate "goodness." Throughout the battle, both sides use morally questionable tactics the Pandavas—through Krishna's guidance (and perhaps manipulation)—violate traditional laws of warfare as understood for that time.

Arjuna defeats Bhishma only through the use of a human shield. While the person is a combatant, Bhishma had sworn never to harm or engage that person in combat. Consequently, Arjuna is protected while showering arrows onto Bhishma. Drona is only overcome after Yudhishthira lies, telling Drona that his son has just been killed on another part of the battlefield. Devastated, Drona cannot continue fighting and gives himself up to death. Krishna, in the heat of battle as Arjuna's chariot driver, urges Arjuna on. Karna lowers his bow and quiver asking for rest, and Arjuna obeys a command from Krishna, firing a volley of arrows into the unarmed Karna.

Duryodhana then takes command and faces off against Bhima. As the two battle, Krishna tells Arjuna that Bhima "will never win in a fair fight [… Bhima] must bend the rules" (Satyamurti 602). Arjuna catches Bhima's eye and "slaps his own thigh" (602). Bhima understands, waits for the opening, and strikes both Duryodhana's thighs with his mace "breaking both instantly" (603). He is later chastised since "All treatises are clear that in a fight, no blow must strike below the belt" (Satyamurti 603). While Krishna inspires Bhima's illegal attack, he then chastises Yudhishthira for not doing anything after Bhima touches fallen Duryodhana's head with his foot while boasting over his fallen cousin. This shows the subtlety of Krishna's understanding of when *dharma* and codes must be bent. Krishna's advice seems to bless the concept of "we did what we had to do," placing it in opposition to Bhishma's living by a code, even if the situation is so extreme that living by that code will mean extinction.

Most loved by all participants in the battle is Bhishma because he is their "affectionate, wise counselor, principal link with the ancestral past" (Satyamurti 466). Having done all he could to avert the war, Bhishma leads the Kaurava forces, yet finds himself fatigued with the world and the battle surrounding him. Bhishma directs his chariot toward Arjuna, assuming a beatific posture, standing "tall, calm and beautiful, hands together […] He was smiling" (Satyamurti 465). Arjuna fires relentlessly at Bhishma, piercing his body with so many arrows that multiple shafts stuck out at every angle. As Bhishma fell his "body did not touch earth but was suspended, as if on a bed" (Satyamurti 466) of arrows. At this all "stood motionless, having no appetite for battle now. Some wept, some fainted, some extolled Bhishma, some cursed the order of kshatriyas" (Satyamurti 464–66).

All lay down their weapons, gathering in silent homage around Bhishma. His head is hanging, uncomfortably unsupported, and Arjuna takes up his bow, consecrated it, and shoots three arrows into the ground, at just the right height to hold up Bhishma's head. Bhishma exclaims happily, "This is a pillow for a warrior! […] I have reached the highest state available to a kshatriya" (Satyamurti 467). Bhishma, mortally wounded but not yet dead, lays suspended between earth and sky. The next day there is no battle as all come forward to pay respects to the great Bhishma. Bhishma lays in state, alive but awaiting the proper moment to take leave of his body. He is attended by human and divine.

After the appropriate time of respect for Bhishma, the war resumes. The Pandavas eventually overcome the Kaurava forces, though Yudhishthira is haunted by the means they employ to defeat Bhishma, Drona, Karna, and Duryodhana. It must be recognized Yudhishthira was not granted the special knowledge of *dharma* Krishna shared with Arjuna prior to battle. The tactics of guile, deceit, and lies used in the war to overcome the Kauravas are in direct contradiction to Yudhishthira's nature, as a son of the god Dharma. Throughout the battle, even Arjuna, knowing the truth Krishna granted, vehemently protests the means used to defeat the larger and more powerful Kaurava forces, especially their commanders. Yet it is Yudhishthira who suffers the greatest moral injury because he holds himself accountable; bearing the condemnation of all of the widows, parents, and families of the deceased on both sides.

Yudhishthira feels that he has betrayed *dharma* and what is "right" for the sake of winning the war. Even though all the cosmic forces conspired to ensure that Yudhishthira prevailed for the greater good of the earth, he cannot accept his actions as justifiable. Krishna tries to comfort Yudhishthira by explaining that there was no other way to win except by employing deception. But the actions advised by Krishna—and taken by the Pandavas to win—weigh heavily on Yudhishthira.

His reaction is self-condemnation, to abdicate his role and responsibilities to those left on earth, and disappear into the forest, living "in the woods […] eating roots and berries, wearing rags, […] harming no one, meditating on the Vedas, […] I shall drift like the wind about the world until the dissolution of this body" (Satyamurti 656–57). Yudhishthira wants to forget who he is and be forgotten by all others.

Yudhishthira sits outside of the gates of the capital city, refusing to enter. He cannot go home. All of the seers, his brothers, mother, wife, and even Krishna himself assure him that what occurred was blessed by and done in the service of the gods. All of these feelings compound to create an immobilizing effect and he has embraced inaction, contrary to his *dharma*. He, physically and spiritually, cannot enter the gates.

Vyasa counsels Yudhishthira, "[You] should be confident […] that you have rightly followed kshatriya dharma. […] Sometimes […] right looks wrong. That's how it is with you. You have acted rightly, as was ordained, yet, blind to this, you burn with wrongheaded guilt. [sic]" (Satyamurti 670). Yudhishthira asks for more guidance and Vyasa encourages him to go to Bhishma, on his bed of arrows. "I am not worthy," said Yudhishthira, "to approach Bhishma, guilty as I am of his great suffering" (Satyamurti 671). To this Krishna admonishes him, and finally Yudhishthira puts "aside his spiritual torment" and finds "some peace of mind" (671) and enters the city, resolved to seek more answers from Bhishma.

Bhishma, on his deathbed of arrows, advises Yudhishthira on how to live a *kshatriya* life. Yudhishthira learns the way in which to move forward. And for Bhishma, his time arrives to leave this plane of existence. Yudhishthira holds a great horse sacrifice ceremony to atone for the war. This ritual allows him and the entire kingdom to move forward.

After living full lives, the time for the Pandavas to go into the wilderness for the final time. Living "austerely, they first turned eastward toward the rising sun and the eastern mountains, following the course of the mighty Ganga to where its waters flow into the sea" (Satyamurti 831). The Pandavas continue into the mountains. One by one, climbing upward, they fall to Time and pass away until, at last, only Yudhishthira, the incarnation of Dharma and the last and eldest of the Pandavas, remains. Much has been put on his shoulders from the time before his birth to now, and he has endured much to serve others. At last, he is met and taken bodily to heaven, escorted by Dharma and Indra.

Yudhishthira reaches heaven to find his earthly enemies, the Kauravas, and all who fought him on the plain of Kurukshetra enjoying peace and abundance, while his own family—the Pandavas and Draupadi—are languishing in pain in a hellish landscape. The images are dispelled as he learns the final lesson, the same lesson Arjuna was given by Krishna just before the battle. Only here, at the end, is the truth revealed that all play their parts as required by the cosmic forces and thus uphold *dharma* in their own way. Yudhishthira learns that in Heaven all hatreds and discords cease. Yudhishthira is led to those he loves, including his cousins the Kauravas, and they embrace in unity.

Martial Culture

Ideally, adherents to martial culture are guided by balanced values, beliefs, and norms. At the core of each martial culture, there is usually a set of principles that forms the bedrock of every decision and action. The ideal of serving something beyond the individual and thereby subsuming the individual to the collective in the form of selfless service is, within US martial culture, a common theme. Those who stray from this ideal to serve more self-centered, secular, and/or personal interests pollute martial cultures. The *Mahabharata* begins generations before the war, describing the reasons why the war was necessary originating with the fall of a golden age, when warrior-priests ensured the cosmic balance on earth through living the ideals of serving something beyond oneself.

Fall of a Golden Age

In the Mahabharata, it should be noted that the *kshatriya* aspect of the individual is passed through the matriarchal rather than the patriarchal line. With this balance, there "was respect for dharma; people behaved harmoniously, without lust or anger. They lived long lives in peace and kindness, free of all disease. Plants and animals also flourished, each in the proper season" (Satyamurti 8). The text indicates the combination of both *kshatriya* and *brahmin* traits creates a balance which in turn balances the earthly realm. The learned/spiritual *brahmin* and warrior/leader *kshatriya* traits combine to create a being of unity in thought and deed. The union of the spiritual and physical allows for selfless action in service to another.

However, sometimes a toxicity can appear and grow in martial cultures. The *kshatriya* of demon stock "despised lawful ways and moderation [and ravaging] the earth, [...] multiplied, sowing hate, mistrust and fear, oppressing and slaughtering gentler creatures" (Satyamurti 9). These oppressors may have been born in the *kshatriya* class, yet they have turned their back on what it means to be warrior in favor of self-interest. At the end of the *Mahabharata*, Yudhishthira proclaims:

> I never give up a person that is terrified, nor one that is devoted to me, nor one that seeks my protection [...] nor one that is afflicted, nor one that has come to me, nor one that is weak in protecting oneself [...] I shall never give up such a one till my own life is at an end. (Ganguli and Vyāsa, *Volume XII, Mahaprasthanika Parva, Section III,* 5)

This statement summarizes what Yudhishthira, the eldest brother of the Pandavas and incarnation of the god, Dharma, sees as the purest ideals of his duty on earth. Yudhishthira's statement stands in direct contrast to the perversion of the *kshatriya* class on earth at the beginning of the *Mahabharata*. Utilitarian attitudes fueled by careerism in military service disregard, starve, and permanently corrupt the human/spiritual ideal of the warrior that seeks to serve something beyond themselves. That secularism, in turn, creates psychological and spiritual loss and damage that echoes throughout the current US martial culture. There are servicemembers and veterans with skills of the warrior who either have never embraced selfless service or have abandoned those ideals. The motivations of a servicemember, like all people, reside internally and can only be perceived through observing the warrior's interactions with others.

Vows

Vows, creeds, and codes are largely informed by honor and shame. As shown in chapter three, current US martial culture seeks to move beyond vows, inculcating the ideal values, beliefs, and norms into the individual. Oaths are taken by both servicemembers in US martial culture as well as those *kshatriya* and warrior *brahmins* within the *Mahabharata*. The oaths taken by those within US martial culture are limiting and do not reflect the reality of service today.

Throughout the *Mahabharata*, the tensions born of the choice between attachments and aversions that interfere with *dharma* such as love, pleasure, envy, happiness, jealousy, and personal peace are a constant theme. Bhishma epitomizes how emotion-laden states need not outweigh the *kshatriya*'s adherence to *dharma*.

US martial culture also recognizes honor and duty must be internalized to be of true value. Bhishma goes beyond extrinsic motivations and internalizes a sense of selfless service. The oft-repeated idiom of "do the right thing, even if no one is looking," is popular among US military leadership, and is meant to reinforce the idea that values, beliefs, and norms must be internalized. The root from which action should emanate is identified through vows and oaths. As Bhishma elucidates:

> "[M]y vow is more important to me than life itself. [...] Dharma for times of distress does not extend to breaking solemn promises," he said. "My vow is everything. The words, once uttered, can never be unsaid without dishonor. This is my truth, and truth for me is greater than all the possible rewards of earth or heaven." (Satyamurti 21–22)

The oath one swears or affirms to enter US martial culture is a brief summation of the servicemember's commitment to service. The seriousness of a service-member's words and actions in martial culture is described by Colonel Robert Dalessandro, "words must be unimpeachable. Oaths are the most sacred promises of all" (Dalessandro 5). Oaths and creeds are, in a sense, mantras designed to return an individual's consciousness to a state of awareness that allows the person to anchor to core values. As Riccardo Rodriguez, a retired US Army officer reflects, "Loyalty, duty, respect, selfless service. I'd like to think I still aspire to live my life in that way" (Rodriguez).

A warrior's deeds must reflect their words; trust is necessary in martial culture and must be rooted in something deeper than externally enforced honor. Which is a virtue informed by external constraints, such as judgment of others by the culture. The spirit informed by the commonality of both the US military en-listed and officer oaths cements the two into a bond of trust and warrior purpose (Swain and Pierce; Dalessandro 6). Both oaths include the sentence: "I will sup-port and defend the Constitution of the United States against all enemies, foreign or domestic, that I will bear true faith and allegiance to the same" (Dalessandro 5). There is an implication that the US Constitution symbolizes an ideal for both a land and a people.

These oaths are less complete than the described duties of a *kshatriya*. For the Pandavas, the cosmic charge is to defend the earth, not just a single people or nation-state. The oath to serve and defend Americans is limiting when one considers the numerous noncombat operations conducted by the US military. Humanitarian deployments to aid in emergency rescue, recovery, and protecting and saving lives have no "enemy, foreign or domestic" but are nonetheless a call to service.

Further, US martial culture's defense of the earth extends beyond the nation's borders and peoples. A recent example is the insistence by senior military lead-ership to provide for the evacuation of Afghans under threat from the Taliban. General Mark Milley, then Chairman of the Joint Chiefs of Staff, stated in May of 2021 "We recognize that a very important task is to ensure that we remain faithful to them, and that we do what's necessary to ensure their protection, and if necessary, get them out of the country, if that's what they want to do" (Copp). The images of US Marines carrying Afghan children to safety in Kabul, Afghanistan, and Air Force crews violating safety regulations to pack just one more Afghan civilian into their aircraft to fly them to safety demonstrate a commitment to *kshatriya* values undiminished by the illusion of "us and them." The US military enlisted and officer oaths to serve the land and peoples of the United States fall short of the reality of what servicemembers are called to do, which transcends the illusion of nation-states, regardless of the intent by the civilian leadership who sends them. The epiphany experienced in true human connection that moves the US soldier to hold an Afghan child breaks through to transcendent principles.

Martial cultural values do not disappear when a servicemember leaves the military. As the situation in Afghanistan deteriorated, many veterans, fearing politics would interfere with the saving of Afghans, took it upon themselves to arrange and conduct evacuations. Several independent but linked groups of volunteers emerged, leveraging social media networks and old contacts to coordinate and evacuate Afghans and their families. These sorts of volunteer organizations populated by veterans have always been in existence "in the background" and range from antipoaching organizations serving in places like Africa to animal rescue organizations who aid active duty servicemembers in bringing beloved animals from war-torn countries to the United States. US martial cultures' values, beliefs, and norms extend beyond the confines of the national oaths taken by officers and enlisted, expanding beyond "US national interests" and seek to serve others of differing cultures.

To bridge the gap between oath and service, US martial cultures develop specific mottos, creeds, and codes to encapsulate the realities of roles in servitude. The creeds of the various services as well as mottos such as the US Army Green Berets "*De Oppresso Liber*" commonly translated as "to liberate the oppressed" serve as supplements, bringing the servicemember far closer to serving those who "seek protection" or "are afflicted" as the *kshatriya* is charged to do.

Oaths are anchoring mantras intended to be taken seriously in martial culture and internalized as guiding principles throughout one's lifetime. US military organizations, often utilized as secular tools of the national government, still understand the true nature of the sacred nature of oaths. Yet this understanding is not universal and has atrophied in many servicemembers. Oaths and vows not only tether *kshatriya* to service but are also taken by those who choose to accompany *kshatriya* and servicemembers into martial culture.

Spousal Sacrifice

Spouses face difficulties in moving from civilian culture to martial culture. To this end, they have few models from the past. The sacrifices of the family members of *kshatriya* demonstrate the path of the *dharma* asked of them. The *Mahabharata* includes accounts of those who are joined to martial culture through marriage.

The sacrifices of Gandhari, Kunti, and Madri symbolize what is asked of people who chose to join their partners in US martial culture. The culture is arguably more alien for the spouse and children of the servicemember than it is for the inducted servicemember. People who choose to marry into or who are born into martial culture do not experience or have a true, formal initiation path that allows for a similar type of acculturation process that the servicemember experiences. This creates, for many, a great deal of psychological and emotional stress, as the spouse and children often do not understand many aspects of martial culture, except through long association. In the absence of a sanctified initiatory threshold experience, many spouses have formed groups which have become formalized. Over the years, spontaneously created unofficial initiations and rules have governed these groups, some of which are perceived less as welcoming rituals and more akin to bullying and extreme hierarchical abuse. The misunderstanding of ritual and initiation demonstrates a necessity for a sanctified integration into martial culture.

Military spouses in today's martial culture often find themselves sacrificing aspects of themselves to martial culture. Like Gandari who binds her eyes to match her husband's blindness, they may have to adopt their spouses conditions of rank and community. Like Kunti and Madri, who voluntarily join their spouses in going into the wilderness, Military spouses often leave behind all familiar things, recognizing the necessity of a holistic joining and welcoming into the culture to balance the self is apparent.

Prism Reflection of a Servicemember

Servicemembers are complex living beings who require a balanced self. The belief that those within the military or veterans only exist within a conflict-seeking persona is incomplete and leads to a misunderstanding of the aspects of a warrior. The Pandavas can be viewed as various aspects of one warrior, while each is formidable on their own and whole only when they act together.

Yudhishthira is the "greatest of the upholders of the Law [...] widely renowned in all three worlds, glorious, lustrous, and moral" (van Buitenen 255). His quality of leadership is a necessity for those in martial culture who must trust each other, and especially those who lead them. Bhima is the aspect of the servicemember that stands for tenacity, courage, and the ability to tap the natural destructiveness within. Arjuna represents the aspect of the servicemember who strives for perfection and excellence in their martial skill. The Asvins represent the healing aspect of the servicemember. The *Vedas* describe the Asvins in a deeper way. It is inferred that Madri's children inherit the Asvins characteristics described in the Vedas as "succouring divinities" who are "the speediest deliverers from distress" (Macdonell 129). The Asvins are "divine physicians, healing diseases […], restoring sight, curing the sick and the maimed" (Macdonell 129). The healer aspect is a recurring theme in martial culture and is recognized as a necessity in treating physical, mental, and spiritual wounds.

The implication of the Asvins as being brothers within the Pandavas as well as fellow warriors and healers leads to the concept of cultural peer or elder led healing. Psychologist Zach Skiles, a Marine Corps veteran who served in Iraq and now one of the leaders in bringing psychedelic healing methodologies to veteran treatment, remarks:

> It's fundamental to our physiology to have peer support as a part of our healing […] there is a theory that emotions themselves have adaptive or cooperative efforts. And so how we interact is actually the basis of how we will heal, [….] when it's about you and your healing there needs to be folks with you who have walked your path and shared your experiences in order to have that affective connection and emotional healing. (Skiles)

The Asvins as peer healers within the Pandavas represents the concept brought out by Dr. Skiles recognizing the power of finding healers within a cultural group. That is not to say that civilians are not capable of successfully working with veterans and their families. Many do so with incredible results. However, those civilians must be aware of the cultural differences that create another layer to work through in terms of creating trust for successful healing.

The Pandavas each represent an aspect of a holistic warrior who guides through wisdom, counsel, and moral virtue, who is capable of great strength and courage, dedicated to skill and removed action, and (represented by the twins) who provides care and exhibits compassion as a healer. While by no means a complete list, these facets come forth as necessary to fulfill the various duties of a warrior.

Toxic Relationships

Toxic interactions are not unusual within military organizations. They contribute to the degradation of martial culture. Individuals who overvalue their personal advancement, or who adapt to a secular, industrialized, bureaucratic system that encourages careerism or personal promotion as equivalent to, or even above, mission success and care of subordinates poison martial culture and promulgate values, beliefs, and norms which are antithetical to martial culture. The *Mahabharata* reveals the dilemma of performing duty under the direction of a toxic leader.

Within the context of the *Mahabharata*, this conflict serves the cosmic and social orders to bring about the war necessary to save the earth. The feud between Duryodhana and the Pandavas is relevant to martial culture in the craving and abuse of positional power, personal advancement, and the placement of careerism over all other considerations.

Toxic leaders generally value their image or persona as viewed by those who supervise them, not those they are in a position of power over, and tend to give favorable ratings to those who are like themselves. Such behavior can lead to preferential treatment and advancement within the military organization of people who are more bureaucratically inclined, less open to critical and creative thinking, and more prone to conform and maintain the status quo rather than mission and team-oriented innovation which inculcates the bedrock cultural values of US martial culture.

The emphasizing of self-serving attributes fosters distrust, envy, and poisonous interactions as a result of objectified valuation. "[You] wind up being micro-managed by people above you because they don't know you well enough, they don't trust you, and people are coming and going constantly" (Ricks and Conan). This misuse and abuse of sacred trust contradicts the ideals of US martial culture. Martial culture demands selflessness rather than seeking power and exercising positional authority for personal gain.

Toxic environments create an inequality in treatment and reciprocity. Toxic behaviors and actions are unbecoming of *kshatriya* or servicemember. Warriors are charged with the care and protection of not only the people, but all within martial culture, whether under their command or not. A way of combating toxicity is to nurture and value select elders within martial culture who embody aspects of the learned warrior. Bhishma is an example of one who inspires positive aspects within martial culture whereas Duryodhana fails in this respect.

Developing Servicemembers

The goal of all training within the US military is to ensure that no matter the conditions under which a servicemember finds themselves, they can fulfill their role within the team. To this end, training and education, accompanied by threshold tests, are vital to the spiritual, physical, and psychological growth of the individual within martial culture. The idealized goal is the ability to find a shared plane of consciousness where a team of servicemembers acts as a single entity, appearing to fulfill their roles effortlessly while adapting quickly to any number of chaotic external stimuli. The training of the Pandavas in particular demonstrates this idealized education. There are several positive examples of the pursuit of excellence within the *Mahabharata*.

Worthiness is a common theme in martial culture and relates to the success or failure of passing a threshold. One must go through a series of austerities before receiving any great knowledge or weapon as either can be dangerous without a properly disciplined and calm mind. Austerities challenge the physical, mental, and spiritual elements of a being. During basic training, just waking up at 4:00 a.m. may be an initial austerity, based upon the recruit's threshold of experiences. Later, as one becomes seasoned and is proven ready to receive more advanced training that will lead to positions of graver danger and greater responsibility, the austerities coalesce into threshold experiences. These experiences increase in intensity thereby providing a scale on which one's worthiness of access to advanced knowledge is measured.

Austerity, or seeking suffering to cultivate skill and obtain greater knowledge or insight, is typical in most martial cultures. Many within the US military "subject themselves to suffering in the broad spectrum [...and] dip their toes in the pool of misery" (Carter) in order to increase their abilities and expand their thresholds, thereby redefining what is true suffering versus mere discomfort. Austerity is a type of suffering that, if accepted willingly, may teach what lies beyond suffering. The military is unique in its ability to create circumstances designed to teach a person "to endure the suffering" (Carter). The various physical, psychological, and spiritual challenges that many servicemembers face in training are a necessity that they, like *kshatriya* youths, must endure before being given the enormous responsibilities incarnate in defending a people through the use of force.

The goal of all martial training is to find a unity of action and intention that is free of distractions, allowing for a purity of movement unhindered by doubts about what is possible. Not every student will find this place within themselves. "A master can only teach a pupil those things he is ready to receive" (Satyamurti 52). The description of Arjuna's realization resulting in his perfect shot on a dark night seems to have the quality of what Abraham Maslow referred to as peak experience.

Maslow posits that the peak experience can be referred to as "a 'little death,' and a rebirth in various senses" (Maslow). Yet it must be recognized that Arjuna is not the most skilled and resolute warrior within the *Mahabharata*, demonstrating that no matter how great one is within their particular sphere of awareness, there is always the unknown. At the end of the great war Arjuna prevails only through skill combined with help from cosmic intervention.

"Unworthy" Servicemembers

Unfortunately, the attitude of determining who should be allowed to serve based on factors other than meritocracy is still prevalent within the military organization today. This is evident in the struggles many have experienced before being allowed to serve, as described in previous chapters. The US government, in very recent times, made a controversial decision to authorize women access to combat-arms roles, both traditionally and legally withheld from women for centuries in most martial cultures. The banning of women from combat was neither wholly successful nor necessarily based on historically accurate norms. The *Mahabharata* demonstrates how prejudice, herein based on social status, pollutes martial culture.

Karna is representative of a *kshatriya* raised in a caste incongruent to his nature and, when he finds his true calling, is rebuffed by the martial culture as unfit.

Karna is the betrayed warrior, a trueborn *kshatriya* who is set adrift in a basket on the river to be rescued by a loving couple who, though lower in the caste order, are open, kind, and generous towards Karna, who they see as a blessing from the gods. Even so, Karna remains adrift without understanding why he is lost.

Historically, governments or rulers have been cautious about who they allow to be taught the application of force, usually out of a fear that those same individuals will turn on them. Examples include the disputes about allowing African-Americans to become combat soldiers during the US Civil War and the introduction of the Women's Army Corps (WAC) in World War II in order to keep women out of the regular armed forces. The United States is no exception to being selective about who should be inducted into martial culture nor slightly wary of all those who have been, regardless of background.

The US martial culture has used idioms like "you are all green" meant to indicate that regardless of race, religion, ethnicity, socioeconomic background, and now gender, the only thing that matters is a person's competence and the ability of the martial-cultural family to depend on the servicemember. This declaration describes the often unrealized ideal of meritocracy as a cornerstone of martial culture and yet it is highly sought after as a state of existence and creates a seldom realized expectation.

In her memoir *Shoot Like a Girl*, Mary Hegar describes an experience she endured as a high school student. Hegar asked a trusted high school teacher for a letter of recommendation to apply for a U.S Navy scholarship to attend college and become a naval officer. The teacher, a US Navy veteran, had been a "big supporter of [Hegar's] dreams" (18). He agreed to write the letter, but instead of praising her potential, the letter was decidedly negative. When Hegar confronted her teacher, he defended his position, "The Navy is no place for you, Mary. What are you trying to prove? This isn't a game. Defending our nation should be left to the strong, and it's no place for a woman" (Hegar 21). Hegar would go on to become a helicopter pilot in the Air National Guard, serve three combat tours, be shot down and—despite her wounds—engage in direct combat to protect her crew and patients. Later, she challenged the US government's Ground Combat Exclusion Policy, which kept women from serving in combat roles. Reflecting on her high school experience, she would "see it for the example it was. Mr. Dewey was simply the first of many people I would soon meet, a faction of American citizens who truly believed they had to protect me (and protect our nation's military) from harm by denying me the opportunity to serve" (Hegar 21).

Both Karna and Hegar face discrimination based upon their station and gender, respectively. Yet both find a way around the barriers placed in front of them. While Hegar served honorably in the US military, Karna's conduct within the *Mahabharata* is often viewed on the whole, as extremely honorable and his skill as superior to the great Arjuna.

Duryodhana, despite usually being caste as the "bad guy," rewards Karna based upon meritocracy and breaks with the caste rules. In the US martial culture this is considered virtuous, whereas in Hindu culture at the time of the writing of the *Mahabharata*, the maintaining of cosmic order is believed to rest in the caste system. Duryodhana's elevation of Karna is, at the time and place of the *Mahabharata*, viewed as a refutation of *dharma*.

The failure of the Pandavas to embrace Karna brings to light one of the many moral conundrums that arise when reading the *Mahabharata*. While the Pandavas are incarnations of deities on earth, they still find themselves trapped within the social context of their births. Caste hierarchy looms large in their estimation of who is worthy and who is not. And their behavior toward Karna demonstrates favoring of caste social rules over ability and meritocracy.

While there are fundamental differences between the Hindu caste system and the US military, both systems are hierarchical and the US military maintains clear divisions between officers, warrant officers, noncommissioned officers, and enlisted servicemembers. Yet there are ways in which one can go from lowest enlisted rank to general officer. Additionally, the division between ranks, especially in more elite units, is significantly blurred. Also, while rare and discouraged, personal relationships between officers and enlisted can occur and are not completely forbidden, unlike the practices of the caste system. Indian culture supported the caste system throughout both civilian and martial cultures in the time of the *Mahabharata*, whereas the US martial culture of today does not extend into the US civilian culture.

The hierarchal divisions between servicemembers within the US military organization can, and often times do, fade when units move from training to actually working in the world. This threshold comes to the Pandavas and Kauravas as well when their period of training and competitions between each other ends. The Bharata cousins must move forward into the world and apply their skills in service to the kingdom. Experience begins to distinguish character as greater than feats of skill ever could.

Preparing for War

To disregard a peaceful life and embrace living in the wilderness is something every servicemember within US martial culture understands. From the outside, it makes little sense to volunteer to place oneself in a situation to be harassed, yelled at, exercised to exhaustion, suffer sleep and food deprivation, and so on to be rewarded with years spent in places devoid of comfort, experiencing extremes of nature, surrounded by danger, and all for little to no financial gain. Eventually, the external motivation provided by the instructors transitions to internal motivation as the servicemember moves to internalize the values of martial culture. This internalization is the activation of the internal warrior, as one acts out of their own center, regardless of the environment and conditions the servicemember finds themselves. The text discusses preparations, which, through hermeneutical examination reveals practices necessary that go beyond training for combat.

To be whole, a warrior must study many different subjects as well as center themselves spiritually. Visiting a favored place prior to stepping on to an aircraft or ship ready to bear the servicemember away is not necessarily universal but is very common in martial culture. The hope that one may come back to the place and/or people they love is a sustaining force. "Yudhishthira toured the streams, crags and copses he had come to love. He looked upward. 'I leave you now,' he said to the silent mountain, 'but when we have regained our stolen kingdom, I shall return to you, as a penitent'" (Satyamurti 263). Yudhishthira's bittersweet farewell to the mountain is understood by many within US martial culture today. A final favorite meal, a visit to a lake, beach, parents, grandparents, or other subtle but substantial rituals/pilgrimages are taken as one prepares to move from the known to the unknown. Jason France, having retired after over 30 years of service, was able to return to his childhood goal of hiking the Pacific Crest Trail and reunite with mountains and forests "as a penitent" (France, Satyamurti 263).

Pleas for Peace

The role of a senior military leader does not only ensure the civilian leadership is able to rely on the military as an organization to be trained, equipped, and prepared, but also to advise against the use of that force (or provide alternatives to using force) whenever possible. This understanding of elders in martial culture as peace-seekers is usually not publicly or popularly understood in civilian or martial cultures. This misunderstanding is demonstrated in Feuerstein's analysis of Yudhishthira's efforts at peace, that his "behavior is more like a *brahmana*'s than a warrior's" (Feuerstein and Feuerstein 32) because he seeks a nonviolent resolution. Yudhishthira is a warrior on the eve of marching to battle who seeks a way toward a peaceful outcome.

In response to the envoy sent by Duryodhana, Krishna's position is that to allow what the dominant part of a culture considers injustice to continue through inaction is a greater sin than what might be incurred through acting to correct the injustice. While this assertion is seen as valid by many, the choice of actions a *kshatriya* has at their disposal range beyond resorting to physical force. Many do not realize that even if militaries are assembled and preparations are made, war can and should be averted by dialogue.

Karna, as the eldest Pandava, is offered the keys to the kingdom, yet refuses. Karna views his adopted parents as his true family and Duryodhana as a true friend. By making the decision to keep his oath (as Bhishma would), he does not reject those he feels beholden to out of convenience. Karna rejects abandoning his honor and those who he loves. His decision, distressing as it fails to stop a war, allows him to avoid moral injury by upholding his *dharma* and the values of the *kshatriya*. However, while Karna's loyalty guides him to uphold his oaths, Karna's choice can also be interpreted as less about keeping one's oath than being blind to greater *kshatriya* values of safeguarding those who seek your protection and looking for ways of peaceful resolution.

Values Betrayed

For those who served in Afghanistan, the withdrawal and evacuation of US forces in 2021 will be remembered with heavy emotions. Those servicemembers must live knowing they had to follow a duty to civilian leadership that resulted in leaving Afghans behind who could not protect themselves. Many in martial culture may find themselves serving leaders who will appear to fail them by betraying what the individual believes are the values of their martial culture. The term in use today to describe this condition is moral injury. Moral injury can happen in many ways, but the two main ones occur when there is a betrayal of what is viewed as right, either by trusted authorities or by the individual themselves. This condition creates cognitive dissonance that ruinates the individual affected.

The fall of the Afghanistan national government and the emergency evacuations that occurred as a result caused many who served in Afghanistan to experience the first kind of moral injury. While withdrawing from Afghanistan is not necessarily considered a betrayal, the way it was conducted and the perceived abandonment of the Afghan people, along with a sense of futility that those who sacrificed so much did so with the hope of believing in the "rightness" of their mission has created a mental and spiritual anguish among those of US and allied martial cultures. Especially among those who personally served in Afghanistan.

While the "president's military advisers argued for keeping those 2,500 troops on the ground to pressure the Taliban to reduce violence and continue peace talks," veteran organizations argued, "We should get out. But this is not the right way to do it" (Bowman). The mental health concern became so widespread that the Veterans Administration, along with other veteran groups, anticipated a deluge of mental health needs: "The moral injury and mental health impact of this withdrawal is huge […;] what was the point of their sacrifice?" (Bowman).

In the *Mahabharata*, on the eve of battle, Bhishma, Drona, and Kripa demonstrate, by their actions, that one may maintain devotion to *dharma* even in the face of difficult situations that pit the values, beliefs, and norms of martial culture against the necessity of obeying the larger civilian culture, which is represented by their elected or chosen leaders. Krishna finds it necessary to teach Arjuna this concept of action in a world free from attachment—seemingly contradictory notions—while maintaining honor within a dishonorable operation or war.

Bhagavad Gita

The experiences of those who are part of the US martial cultures who have served since the dissolution of the draft have, especially after 2001, experienced a lifeway of endemic conflict. Multiple deployments that are broken up by coming back to their installations for "normal" life create a lens of the world as constantly suffering conflict. Yet this does not stop the servicemember from "going out" again. At some point, however, there may come a moment when one asks, "What is the point?" As Chaplain Pratima Dharm—the Pentagon's first Hindu and interfaith chaplain—describes, within the *Bhagavad Gita* the basis for war "is explained by the words of Shri Krishna to Arjuna trying to bolster him [...] to see the real from the unreal [the *Gita*] is a guide to us [about] how we live in this world—we are called to do our duty, we are called to find a purpose in this world" (Lakshman). Dharm used the *Bhagavad Gita* as her basis for counseling US soldiers of Hindu faith who were deployed in Iraq. Her use of the Hindu mythos demonstrates the immediate applicability, resonance, and need of US martial culture. Within the *Mahabharata*, the famous *Bhagavad Gita* provides an answer to the question of how to fulfill one's duty.

Warriors can find themselves in opposition to one another through accidents of birth, born within a set of circumstances, defined by parentage, borders, ideologies, religions, socioeconomic factors—all outside of one's control. Despite these circumstances, one must find their true self and be obedient to the demands of that duty. As Dharm describes "It is [...] part of the Bhagwat Gita—how we make sense of war and what is your duty towards that [....] For me, that was my basis, to go to war and to be able to offer the best of myself" (Carroll).

Arjuna has fought other peoples and demons who have opposed his family and community. In this battle, the enemy is now friends and family from his own community. The foe is suddenly humanized in a way Arjuna cannot reconcile with his *dharma*. Because of the recognition of the other as related to self, the experience of dissonance brings the warrior to his knees. Abandoning personal needs and desires in the service of a greater power, such as the infinite root of all creative power (*Brahman*), is the way toward enlightenment and release in the Hindu religion. The overriding message to transcend the self and let go of pursuing pleasure and self-comfort runs throughout the *Gita*. Arjuna worries for his future happiness, anticipating a lifetime of sorrow. Krishna responds that Arjuna is selfishly placing his future personal happiness and peace above the needs of others. Personal happiness is not a concern to one who understands *atman*, the universal Self that is unchanged by experience. Nor should it be the goal. It is how you comport yourself in life that matters most.

One of the aspects of "selfless service" that is not nearly as acknowledged as it should be is that entering into martial culture quite possibly means making hard choices, living with the reality that there are no good options, just less terrible choices. However, the idea of public martyrdom, which is a persona sometimes adopted by veterans, is antithetical to selfless service when one makes a display of their suffering for personal gain.

Krishna's admonishments to Arjuna to abandon ego, pick up the bow, and stand are sometimes necessary for servicemembers who, knowing what war is, understand the heavy price that will be exacted for committing once again to battle. The *kshatriya*, by following the law of action and conducting their *dharma*, provide for and protect the space necessary for those following different lifeways to exist and grow.

Joining a martial culture does not imply a pleasurable life filled with earthly blessings. Living means embodying, not in a symbolic esoteric way, but in a very real and way, the intense memories of the highs and lows of service. Arjuna does not seek to avoid death, but the pains of life. The Mahabharata suggests that the living forward, bearing the memories, is an inherent part of the service one provides within a martial lifeway.

What We Had to Do

There are many codes of conduct about how to comport oneself during various stages of life within martial culture. The reality is that to live, much less to survive in combat, is less about fairness and more about survival and overcoming an adversary. Rules of engagement and international laws have legitimized codes of conduct in battle. Within the *Mahabharata*, facing uneven odds, Krishna repeatedly counsels the Pandavas to violate accepted norms of conduct.

The second definition of moral injury describes the condition when an individual has done something to violate their own internal moral or ethical compass becomes evident here. The actions taken by those who have seen or supported combat operations within US martial culture may cross a line within an individual's consciousness. How to work through that trauma is what Yudhishthira faces at the end of the war. But the dust must settle first.

Bed of a Kshatriya

The necessity to provide a sacred space to mourn and remember those who have fallen has manifested itself in the US military with several personal rituals, but the emergence of officially sanctioned rituals, such as the ramp ceremony, has come into being. These ceremonies represent a breaking through of the sacred, putting servicemembers in touch with the depths of martial culture. The necessity to still the battlefield to offer respect for those who have passed is found within the *Mahabharata*. Bhishma's fall in the war moves all armies to cease hostilities, grieve, and demonstrate respect for his sacrifice. The shock of losing a beloved sister or brother, on or off the battlefield, can drive one to tears, rage, shock, and even despair.

During the US wars in Iraq and Afghanistan, when a servicemember was confirmed killed, either on the battlefield or having succumbed to wounds on the flight back (usually by helicopter), servicemembers throughout the post or base to which the aircraft was going were notified. Those available would gather at the location where the aircraft landed, to bear witness and honor those now "coming home." For a place like a combat outpost, this ritual is small, usually unorganized in appearance, and very personal. At larger airbases where one enters and leaves a country, the ritual is more elaborate. In 2010, Michael Holmes, a reporter at Kandahar Air Base, Afghanistan noticed several hundred servicemembers from many different countries gathering on the airfield.

> They were here out of respect, solidarity with a comrade. [...] Eight soldiers lifted their fallen comrade off the vehicle, another soldier in the front and rear to begin the solemn march to the giant plane's ramp. [...] As the ramp lifted, every soldier was saluting. (Holmes)

In US martial culture, other memorial ceremonies happen on the battlefield shortly after a loss, creating yet another sacred space to allow for honor, grief, and reflection by those within the unit who lost a warrior. Frequently, in Iraq and Afghanistan, the memorial was centered on what is commonly known as the battlefield cross. Originally used to mark shallow graves for retrieval and delivery home later in World War I, this symbol, "rifle, helmet, boots, and dog tags, has become the symbol of loss, of mourning and closure for the living" (Golden).[6] The helmet rests on the rifle, bayonet in the ground, and empty boots standing at attention.

> "There was one guy [...] who died at Abu Ghraib prison [...] when he passed, we had a ceremony for him. You know with the rifle and their helmet and boots. His favorite song was Freebird by Leonard Skinner. And so we stayed in formation at attention for that whole 11 minute song." (Osuna)

Compare the image of the battlefield cross with Bhishma's head resting on three arrows, and points driven into the ground. The synchronicity of these two images draws sacred connections across time and space.

The sacred space necessary for grieving should be maintained and encouraged. In some cases, when under fire, the mourning must be postponed, or others will be lost. The evolution of US martial culture reintroduces this ritual and allows for the liminality to unfold directly, reflecting back to some of the oldest mythologies.

"Peace"

After a battle or the larger war ends there is the peace that follows when the flurry of activity halts. For those who took part, the resulting peace may sometimes be more unbearable than the war. Within the *Mahabharata*, Yudhishthira, in particular, speaks to the concepts of moral injury.

"The gist of Krishna's teachings [in the Bhagavad Gita] is that when the moral and spiritual welfare of a people are at stake, war is permissible" (Feuerstein and Feuerstein x). Given Krishna's advice to Yudhishthira throughout the battle, many things not normally permissible even in battle are allowed when *dharma* is threatened. Many read the *Bhagavad Gita* as a separate text, removed from the context of the *Mahabharata*. Which can result in a different interpretation of Krishna's teachings in light of his application of these words throughout the coming battle and life after.

Krishna in the *Bhagavad Gita* must be balanced against Krishna in the *Mahabharata*. He is not two separate entities but one who is attempting to show that cosmic purpose is beyond human comprehension and, when applied within the world, bears a subtlety difficult to understand. This perspective is difficult to even contemplate, let alone accept. To see oneself and one's adversary both as necessary components of an interactive world and still seek to participate in the battle, even to the point of violating rules of war, seems incomprehensible. There is a paradox that people who are not inducted into martial culture have difficulty resolving is: how can one use force while maintaining empathy, even for the ones you are using that force against?

Yudhishthira suffers from moral injury and cannot accept the declaration of a "just war" when, even if the actions were in accordance with the gods, his own internal feelings pollute the rightness of the actions. The haunting specter of past perceived failure, whether it is true or not, resonates with those in US martial culture today. Hegar discusses coming home and what she carried:

> When we got back, I took some time off [...] I would sometimes wake up in the middle of the night in a sweat. I wasn't dreaming about getting shot down, though. For me the thing that was difficult to recover from was the actual medevac missions and the fact that we couldn't save everyone. I can barely recall any of the times we saved people, which was more often than not, but when I look back at my missions, it's the ones I lost that usually come to mind. Did I do everything I could? Could I have gotten to them faster?. (Hegar 260)

It may seem surprising that being shot down and engaged in direct combat is not the memory that lingers. To those within martial culture who understand the values, beliefs, and norms of placing the team above oneself.

Vyasa and Krishna, in their advising of Yudhishthira, have the gift of time and experiential understanding. In US martial culture, those who are separated or retired are usually not given a space to offer the next generations their wisdom, understanding, and hope. Some retired senior military leaders tend to do this for others, serving as advisors after they have left the military organization. Little mentorship exists for those who leave the military organization and continue to exist in martial culture, inhibiting the continuation of the martial maturation process.

There are many Veteran Service Organizations, yet these largely focus on midwifing newly separated servicemembers into becoming veterans. US martial culture needs to break with the secular, corporate understanding that service within the military is a "job," and provide a space for their own elders internally, shaman-types to provide guidance and, in the crucial time of crisis Yudhishthira currently occupies, provide living proof that one can come home.

Sanctifying the Future

The passing of Bhishma and subsequent funeral rites serve as a poignant change in the world of the *Mahabharata* and recognize the necessity of cleansing ceremonies to close an age of warfare and begin an age of renewal. There is an acceptance that once the knowledge and advice of one generation has been granted to the new, it is time for an age to pass.

Cleansing ceremonies are not really recognized as necessary within the secular US military organization but the need is recognized by those who make up the martial culture within that secular organization:

> After spending four months covered in sand and dirt in the heat of the desert, suddenly the thought of jumping into the Indian Ocean was the greatest idea any of us had ever heard. [...] The ocean was warm and welcoming; I never wanted to get out again. That swim, which I'll remember forever, was both physically and emotionally cleansing. It should be a required activity for every person reintegrating into civilization straight from the violence and horror of the battlefield. (Hegar 59)

Jason France's journey on the Pacific Crest Trail hit a turning point five weeks in:

> "I would just stop a lot, taking in the views. Absorb the sounds, the sights, the smells [....] It took a while for my mind to slow down. [....] There was a day at 10,000 feet and I just felt that click [....] where my heart rate, my respiration, my perspiration, my energy expenditure, just everything seemed to come together." (France)

The current secularized way in which military organizations exist has yet to fully acknowledge the lived experiences of those within US martial culture and the needs that are not being met. Bringing forward long forgotten rituals may be the way in which some of the deepest needs of servicemembers and veterans can be addressed.

Time, Enemies, and Friends

The nation-state's self-interest is evident in the way the US military is employed. However, that interest is sometimes contradictory to the values of martial culture. Even those who join the US military for reasons other than service will find that once they are "on the ground" and see people of a different culture who need protecting, the illusions of separateness and "other" fade and the servicemember, following *kshatriya* warrior ideals, takes under their protection the Iraqi family or Afghanistan child or stray animal who seeks sanctuary, protection, and compassion.

One of the most difficult challenges that occurs within martial culture is how to reconcile and make peace with a once-enemy. In modern societies still ruled by blood-relation lineage, blood feuds between families, clans, and tribes exist for generations. In US martial culture, a servicemember may face an other as an enemy one day, and yet a year later, find themselves facing a different other as an enemy. The advantage, if there is such a thing, of having many adversaries rather than a single life-long enemy, is that it illuminates that the concept of "enemy" is an illusion based upon a series of conditions that ebb and flow with time. Major General Hal Moore commanded US forces in the Battle of Ia Drang in 1965 against the opposing North Vietnamese forces commanded by Lt. General Nguyen Huu An. In a speech, Moore described how fiercely the two forces fought each other and his recognition of the aftermath:

> we dropped to our knees […] It was over. I looked around at the devastation, the dead [….] We had won […] But at what sacrifice? Seventy-nine of my dear troopers died for those of us who lived. […] I witnessed over 600 hundred enemy bodies strewn over the valley. (Moore)

Time is the trickster that reveals the impermanence of "friend" and "enemy" and causes all such labels to fall away. In the Mahabharata, Yudishthira's final lesson is when "residing in Heaven, all enmities cease" (Ganguli and Vyāsa *Volume XII, Swargarohanika Parva, Section I* 1). This is a gift that time can grant if the individual is open to understanding the greater scope of the cosmos.

Soldiers who participated in the invasion of Iraq in 2003 returned in 2006 to train the new Iraqi military which may be composed of the very same individuals the servicemember fought against. Germans and Americans now freely marry, despite being fierce enemies nearly 80 years before. The same is true of the Japanese, who today are staunch allies with the United States. In 1993, Moore and An, along with veterans from both sides of the battle, returned to the Ia Drang valley to meet in peace:

> Nguyen Huu An and I came face to face. Instead of charging one another with bayonets, we mutually offered open arms. [We formed] a circle with arms extended around each other's shoulders [….] With prayer and tears […] we quickly learned that the soul requires no interpreter. (Moore)

In that valley of a long-ago battlefield, where the "soul requires no interpreter," it is as if, like Yudhishthira, these former soldiers were invited to bathe in the Ganga by Lord Dharma where "all resentment, grief, hostility" (Satyamurti 839) falls away.

Conclusion

The *Mahabharata* details life's journey of a kshatriya from before the womb to beyond the tomb; they discovered joy and pain, revealed better and worse natures, and exhibited human frailties and strengths in equal measure in the most extreme of circumstances. The Hindu epic has been a beacon of light and hope throughout the world for centuries. The way in which Vyasa weaved in the very human struggles to understand and live as a warrior throughout a lifetime knew well of what they spoke.

Servicemembers, veterans, and their families could learn much from the *Mahabharata*. The epic tells of how to live a martial lifeway under difficult and confusing circumstances. As Dharm posits regarding her understanding of the experience of war as servicemember based on the Vaishnava tradition:

> War, similarly, is very much a part of this world […] But really, the guiding principle is to live in this world as if things are temporary. [….] That basis is so much present in the words of Lord Shri Krishna. [….] That's what I gave to the soldiers. (Lakshman)

Yudhishthira, Bhishma, Arjuna, and all of those within the *Mahabharata* have their counterparts in every generation. In exploring correlations between this ancient text and the experiences of servicemembers today, many significant revelations emerge. Perhaps one of the greatest lessons passed forward in the text is that warriors standing across the battlefield from one another, having lived the martial life, are often kin separated by circumstance. Servicemembers approaching each potential conflict must attempt to avert war as if battling against a sibling, parent, or loved companion. Should it come to force, be quick to resolve the conflict with all haste while demonstrating compassion to your enemy, conscious that the apparent separateness is an illusion. While a kshatriya must choose action in the world in which all entities exist, one must be mindful of the illusion. US martial culture needs to learn from the *Mahabharata* and apply the teachings it offers before, during, and after all of the sounds of battle have faded.

> "It is said that the day's sins may be dissolved by listening to a part of it at night in a joyful spirit, with a trustful heart, with a perfect quality of attention [….] The Mahabharata is a fathomless mine of wisdom, precious gems of knowledge for anyone receptive to the truth." (Satyamurti 843)

This living mythos exists as a message for US servicemembers and veterans on how to express duty and live life.

Notes

1 According to the Law Code of Manu, the *Brahmin* are the priestly class, the *Kshatriya* are warriors and rulers, the *Vaisya* are merchants and craftspeople, and the

Sudra—commonly referred to as untouchables—deal with the dead and other tasks, such as leatherwork (Olivelle 19; Stein 50–52).
2 Dharma emanates from the cosmic source as the eternal principles that underly the cosmos. These are reflected in guidelines for an entities proper role within the universe. Commonly, it is referred to as duty, however, it is an all-encompassing aspect that dictates one's entire life.
3 Bhishma means "Awesome One."
4 God of Wind
5 A "god of the thunderstorm who vanquishes the demons of drought or darkness and sets free the waters or wins the light. He is secondarily the god of battle" usually recognized with his weapon, the thunderbolt (Macdonell 42).
6 Golden does point out that although it is called a "cross," there is no "overt religious context" (Golden).

References

Bowman, Tom. "With No Parades and Little Ceremony, America's Longest War Draws to a Close." *NPR*, 3 July 2021. www.npr.org/2021/07/03/1012490524/with-no-parades-and-little-ceremony-americas-longest-war-draws-to-a-close

Campbell, Joseph, and Bill Moyers. "Ep. 2: Joseph Campbell and the Power of Myth – 'The Message of the Myth'." *BillMoyers.com*, 19 Oct. 2020, https://billmoyers.com/content/ep-2-joseph-campbell-and-the-power-of-myth-the-message-of-the-myth/

Carroll, Chris. "Military's First Hindu Chaplain Brings a Diverse Background." *Stars and Stripes*, 2 June 2011. www.stripes.com/news/military-s-first-hindu-chaplain-brings-a-diverse-background-1.145455

Carter, Clint. "Inside America's Toughest Survival School." *Men's Journal*, 12 Aug. 2019. www.mensjournal.com/adventure/inside-air-force-survival-training

Copp, Tara. "US Planning to Evacuate Afghan Interpreters, Top US General Says." *Defense One*, 27 May 2021. www.defenseone.com/policy/2021/05/us-planning-evacuate-afghan-interpreters-top-us-general-says/174337/

Dalessandro, Robert J. *Army Officer's Guide*. 51st ed., Stackpole Books, 2009.

Feuerstein, Georg, and Brenda Feuerstein. *The Bhagavad-Gita: A New Translation*. Shambhala, 2014.

France, Jason. Personal Interview. 28 June 2023.

Gandhi, Mahatma, and Mahadev H. Desai. *The Gospel of Selfless Action, or, the Gita According to Gandhi*. Navajivan Publishing House, 2019.

Ganguli, Kisari Mohan, and Vyāsa. *The Mahabharata of Krishna Dwaipayana Vyasa (Volumes I–XII)*. Munshiram Manoharlal Publishers Pvt. Ltd., 1993.

Golden, Kathleen. "The Battlefield Cross." *National Museum of American History*, 21 May 2015, https://americanhistory.si.edu/blog/battlefield-cross

Holmes, Michael. "CNN Reporter Witnesses Solemn 'Ramp Ceremony.'" *CNN*, 23 Apr. 2010, https://web.archive.org/web/20230202170448/http://www.afghanistan.blogs.cnn.com/2010/04/23/cnn-reporter-witnesses-rare-ramp-ceremony/

Lakshman, Narayan. "Gita My Basis for Counselling Hindus in U.S. Military." *The Hindu*, 12 Apr. 2016, www.thehindu.com/opinion/interview/gita-my-basis-for-counselling-hindus-in-us-military/article6605265.ece

Macdonell, Arthur Anthony. *A Vedic Reader for Students*. Oxford UP, 1917.

Maslow, Abraham H. "Religions, Values, and Peak-Experiences." *Bahaistudies*. www.bahaistudies.net/asma/peak_experiences.pdf, accessed 6 Feb. 2021.

Moore, Hal. "From 'Face-to-Face' Combat to 'Arm-in-Arm' Friendship a Speech Delivered by Lt. General Hal G. Moore." *From Dana's Guests*, 23 May 2008, www.danaroc.com/guests_genhalmoore_052509.html

Osuna, Freddy. Personal Interview. 14 June 2023.

Ricks, Thomas, and Neal Conan. "Tracing Military Failures, Holding 'the Generals' Accountable." *NPR*, 11 Dec. 2012, www.npr.org/transcripts/166963643

Rodriguez, Riccardo. Personal Interview. 16 June 2023.

Satyamurti, Carole. *Mahabharata: A Modern Retelling*. W.W. Norton & Company, 2015.

Skiles, Zach. Personal Interview. 25 July 2023.

Stein, Burton. *A History of India*. 2nd ed., Edited by David Arnold, Blackwell Publishers Ltd, 2010.

Swain, Richard M., and Albert C. Pierce. "The Armed Forces Officer - Chapter 1: The Commission and the Oath." *National Defense UP*, 17 Apr. 2017, https://ndupress.ndu.edu/Publications/Books/Armed-Forces-Officer/Article/1153505/chapter-1-the-commission-and-the-oath/

Van Buitenen, J. A. B. *The Mahabharata: 1 The Book of the Beginning*. U of Chicago P, 1983.

Venugopal, Vasudha. "Mahabharata: Mahabharata Much Older, Say ASI Archaeologists." *The Economic Times,* 20 Oct. 2019, https://economictimes.indiatimes.com/news/politics-and-nation/mahabharata-much-older-say-asi-archaeologists/articleshow/71658119.cms

5   Norse

The Norse mythologies hold a great deal of fascination for many people. Gods and giants engage in sometimes humorous adventures that usually include an undertone of urgency as the final battle, *Ragnarök*, approaches. The battle's outcome is already predetermined, each combatant's fate known, down to how many steps Thor will take before succumbing to his wounds. And yet the goddesses and gods fill their days with living, and even, in Odin's case, preparing for the war with a fervor, acting as if the predestined outcome might be overturned. The seeming contradiction between being at the mercy of the *Norns*, three supernatural feminine beings who determine and then weave the destiny of each life, and the struggle by the chief God Odin to overturn the fate of his age encapsulates the paradoxical outlook of the Norse. Yet, the true nature of the beliefs of the peoples of the lands surrounded by the Baltic and North Atlantic seas are even more complex, and their mythological understandings are perhaps more nuanced than is commonly understood. The same can be said for Scandinavian views on martial culture, as they appear in Norse mythologies, rituals, and practices.

Neil Price notes that the way the Scandinavian peoples framed their understanding of body versus soul has much to do with how they interacted with their environment. Both their interior and exterior lives emphasized a comfort with liminality, betweenness, and plurality. The Norse "world hummed with life, but its boundaries, both internal and external, were in many senses more permeable than ours, always and constantly connected by winding paths to the realms of the gods and other powers" (Price 10). Through that interaction with the world, Price asserts that at "the most fundamental level of all, inside every Viking-Age person was not just some abstract 'soul' (if that is to your spiritual taste) but several separate and even independent beings. Each was a component of the whole individual" (Price 32). Therefore, to settle upon one aspect of the self as explanatory of how the Norse person viewed themselves is to promote a colonialist mindset.

There is a parallel between the stereotyping and misunderstanding of the Norse and the modern US martial culture. If the myths are read without incorporating current understandings of historical Scandinavian lifeways, then the

DOI: 10.4324/9781032613222-5

words confirm popular cultural stereotypes. Recent archeological and genomic discoveries, combined with interdisciplinary academic methods, give the myths a wholly different reading and a far more interesting and multifaceted understanding that challenges stereotypes by revealing the natures of the Scandinavians and their goddesses and gods. For example, a genomic study of "442 humans from archaeological sites across Europe and Greenland" published in 2020, noted "'Viking' identity was not limited to individuals of Scandinavian genetic ancestry" (Margaryan et al. 390–91). To be of the Norse culture was not necessarily to be genetically Scandinavian. The diversity of peoples who intermingled to become the Norse perhaps explains the rich, but often forgotten themes of their mythology.

The primary texts of Norse mythologies were made by peoples from paradigms entirely different from, and outside of, the pre-Christian Northern European peoples. Northern European religions were so successfully erased that the written records currently available were recorded by outsiders who observed, interacted, or found themselves in conflict with the Northern peoples (Davidson *Gods and Myths of Northern Europe* 14–15). The first religions native to Eurasia and Northern Africa—what the post-modern world now calls mythologies— were destroyed and demonized by monotheists. Consequently, any exploration of texts must acknowledge the reality of "prejudice, misinterpretation, or deliberate editing when non-Christian beliefs are being dealt with" (Davidson *Gods and Myths of Northern Europe* 15). Even the Scandinavian mythological poems that have survived were recorded hundreds of years after the Icelandic conversion to Christianity. Further, descriptions and observations of the Scandinavians and their cultural practices, such as the Muslim Ibn Fadlan's or the Roman Tacitus's recordings of their interactions with the Northmen or Danes (as they were called then), are also made from outside the cultural context in which the Scandinavian peoples and their ways could be most fully understood. Thus, the study of Norse myths requires inclusion of various academic disciplines because the Norse peoples themselves do not have a preserved literary or oral tradition that reaches through time.

Archeology is providing new information that helps to reframe the commonly accepted interpretations of the *Poetic Edda* (also known as the *Elder Edda*) by Saemund Sigfusson and the *Prose* (also known as the *Younger*) *Edda* by Snorri Sturluson. The growing amount of data being discovered adds depth to the written records and places them in a more accurate context. This data allows for previous understandings of Norse myth to be supplanted by contextualization in a way which was not possible previously.

The current view of Norse mythology is captured, however incompletely, within the *Poetic* and *Prose Eddas*. These compilations of individual poems and stories are derived from older fragmentary codices and oral traditions. Both *Eddas* have been assembled in a way that presents the illusion of a single mythos representative of peoples from Iceland to the Baltic, Kyiv, and the Byzantine empire.

The Poetic Edda was compiled from various preexisting texts by Saemund Sigfusson, a well-known Icelandic scholar of the twelfth century. Snorri Sturluson, drawing drew from the same material, composed to compose the *Prose Edda*, though with substantial alterations to place the heathen Scandinavian traditions within accepted Christian, Greek, and Roman mythological traditions. It is likely the geographic isolation of various farming communities throughout these regions, caused by terrain and weather, likely gave rise to a multiplicity of mythological traditions. Both Eddas hint at a diverse, rich wellspring of mythologies and rituals that are mostly lost to history; and with them the voices of the people who lived them. However, the two Eddas still provide ample mythology to study through a martial cultural lens because even with all of the limitations explained, they are the texts available to us today.

The Norse peoples are brought into greater relief and find themselves given a place within the histories of northern Europe that humanizes rather than demonizes them. The Norse "had courage, vigour [sic], and enthusiasm, an intense loyalty to kindred and leaders [....] They were great individualists [....][...,] capable of considerable self-discipline, and could accept adversity cheerfully without whining or self-pity" (Davidson *Gods and Myths of Northern Europe* 10). This description of the attributes of the Norse, generous as it may be, is inclusive of traits associated with happy warriors who embrace both ecstasy and suffering with equal zeal. The Norse philosophy is an ideal of what servicemembers—especially those going through advanced training—strive toward.

The Myths

The Norse peoples are the descendants of the first two humans, Ask and Embla. Three gods, Odin, Ve, and Vili, were rambling along the seashore and came across two logs. They decided to create humans from these logs. Odin breathed into them life, Ve granted them intellect and mobility, while Vili provided their forms and ability to speak, hear, and see. The gods clothed them and gave the two names. "The man was called Ask [Ash Tree] and the woman, Embla [Elm or Vine]." (Sturluson 18) From these two are descended all peoples.

Ask and Embla are given the realm of *Midgard*, separate from other lands such as *Asgard*, the home of the goddesses and gods. The centrality of the tree sustaining life is seen in the world tree, an ash named *Yggdrasil*, which connects the various realms. The Tree is the giver and sustainer of all things, though it is tortured by snakes gnawing at its roots and those who live off its sustenance.

There are two main groups of deities, the *Aesir* and the *Vanir*. The *Aesir* includes Frigg, Odin, Heimdall, Thor, and Baldur while the *Vanir* includes Frey, Freyrer, and Njord. The *Vanir* "are first and foremost rich and givers of riches; they are patrons of fecundity and of pleasure (Frey, Freya), also of peace (Frey); and they are associated [...] with the earth that produces crops (Njord, Frey), and with the sea that enriches its sailors (Njord)" (Dumézil *Gods of the Ancient Northmen*

4). It is generally accepted that the *Aesir* "determined the course of war and sent thunder from heaven [and] were essentially powers of destruction, although they might also give protection from chaos and disorder" (Davidson *Gods and Myths of Northern Europe* 92).

After a war between these two groups, a peace is negotiated. "They reconciled their differences by the following procedure: both sides went to a vat and spat into it" (Sturluson 84). To preserve this sign of peace, the gods created the wise man Kvasir from the concoction. He brought knowledge to the world of men. However, two dwarves tricked and killed Kvasir. They mixed his blood with honey and created a mead that gave the one who drank it either the gift of poetry or scholarship.

The *Aesir* and *Vanir* live in *Asgard*, a realm of the deities who each live in their own hall.[1] Each deity's hall and farmstead reflected their interests and how the people perceived them. As an example, the land of "Gladsheim, where gold-bright, wide Valhalla stands" (Crawford *The Poetic Edda* 62) is the land of Odin, where warriors train for the great war of Ragnarök. Yet his land and hall is but one of many. Further, the narrative that all warriors who die go to Valhalla is not born out by the myths. Freya of the Vanir, "the most splendid of the goddesses" (Sturluson 35) chooses half of those warriors slain in battle whenever "she rides into battle" (Sturluson 35) and they go to "Folkvangar [Warriors' Fields]" (Sturluson 35). Her hall "Sessrumnir [With Many Seats]" (Sturluson 35) exists within this "ninth land" (Crawford *The Poetic Edda* 63). Odin receives the other half of the fallen, and fills Valhalla with warriors he believes will be of use to him in *Ragnarök*, the final battle already foretold. Here, the *valkyrie* serve those chosen warriors, collectively known as *einherjar*, at their feasting. However, the *Valkyrie*, meaning "she who chooses the slain," weave the fates of those engaged in battle, deciding who dies and lives, and then ferry the slain to Odin's hall, *Valhöll. Valkyrie* are fierce in their presence on the fields of battle and ever ready to elevate or condemn those who muster on battlefields across the earth.

Odin seeks great kings and warriors for his hall, reaping them from Midgard (Earth). Meanwhile Thor, strongest of the gods and goddesses, protects the realms from *jotun* (giants) and other threats. Each has a different focus, Odin the end war of Ragnarök, Thor the ever-present threats. Odin prepares those warriors he favors with grants of boons. One such warrior, a descendant of *jotun*, Starkad, finds himself blessed by Odin. However, as Starkad was descended of giants, Thor seeks to undermine him.

Odin ordains "that he shall have the best of weapons and clothing.

Thor: 'I ordain that he shall have neither land nor estates.'
Odin: 'I give him this, that he shall have great riches.'
Thor: 'I lay this curse on him, that he shall never be satisfied with what he has.'
Odin: 'I give him victory and fame in every battle.'

> *Thor:* 'I lay this curse on him, that in every battle he shall be sorely wounded.'
> *Odin:* 'I give him the art of poetry, so that he shall compose verses as fast as he can speak.'
> *Thor:* 'He shall never remember afterwards what he composes.'
> *Odin:* 'I ordain that he shall be most highly thought of by all the noblest and the best.'
> *Thor:* 'The common people shall hate him every one.'"
>
> (Kershaw 45)

Starkad lives three lifetimes, fulfilling all the positive and negative foretellings of his life.

At another time, Thor and some companions find themselves making a journey to the home of Utgard-Loki in the land of the giants. During this journey, and as guests of Utgard-Loki, Thor and his companions are subjected to a series of tests. Throughout a series of ordeals, Thor in particular is ridiculed for being weak, powerless, and unable to accomplish seemingly minor tasks such as outwrestling an elderly woman. The god to whom the Norse give offerings for deliverance from the dangers of the world is completely humbled in the land of giants. Thor himself doubts his own abilities. Yet after each perceived failure, he rallies to take the next test. Thor keeps struggling until Ut-garda-Loki puts a stop to the tests.

Ut-garda-Loki asks Thor "how he thought the trip had gone [....] Thor replied that he could not deny that he had been seriously dishonoured [sic] in their encounter: 'Moreover, I know that you will say that I am a person of little account'" (Sturluson 61). Thor is mocked over and over again for failing. But the true test lies not in his success, but in how he handles defeat. These are not truly tests of strength or endurance but of mental and emotional strength and resilience. Never does it occur to him to turn down a challenge, quit, or succumb to despair. He is resolute in his attempts, straining every last bit of self. Even though Thor is ridiculed and is convinced he has failed, he continues to try.

After Thor and his companions are outside the fortress, Ut-Garda Loki reveals the truth that he is both Skyrmir and Ut-Garda Loki and explains that each test was an illusion. Where Thor was challenged to empty a horn of ale, he was really attempting to drink the entire ocean. When he thought he was so weak he couldn't lift a cat, it was really the Midgard Serpent "which encircles all lands, and from head to tail its length is just enough to round the earth. But you pulled him up so high that he almost reached the sky" (Sturluson 62). And finally, when he failed to subdue an old woman, he was really wrestling Old Age, which cripples everyone. After this reveal, Ut-Gard Loki, in order to safeguard himself from Thor, disappears both himself and his home. Thor and his companions return to Asgard. Tests, and failure, are vital to finding thresholds and pushing past them successfully. The greater the testing, the greater are the demands of sacrifice.

Access to powerful knowledge requires sacrifice that is sometimes quite brutal and demanding. Odin, always seeking more knowledge, finds that to access the power of runes, he must hang on the world tree Yggdrasil. Pierced by a spear, he hangs in agony. Suddenly, he peers down at the runes and they reveal themselves to him. With a cry, he falls, finally obtaining the hidden knowledge. As the god of many names, Odin demands of himself his own pain and anguish to pass the threshold and be worthy to take up and understand the runes. Odin is known for his dedication to the sword and spear as well as ecstatic knowledge. In this way, two groups of human warriors are associated with Odin, a god who combines ecstasy, hidden knowledge, and a singular focus on preparing for the great battle. The berserkers and Odinic warrior-heroes appear on earth, blessed with the One-Eyed God's powers.

Ragnarök unfolds as foretold. Despite Odin's best efforts, the old world is overwhelmed. Yet there is a rebirth from the ashes left by the last world. The children of the gods and goddesses survive. The sons of Thor, Modi and Magni, inherit *Mjollnir.* The Daughter of the Sun continues her mother's work. And Baldur, the perfect child of Frigg and Odin who died prematurely long ago, returns from the dead to lead the universe into a new age. A woman and man who had hidden in the boughs of Yggdrasil from the conflagration, emerge to find a renewed world. From them descend all the peoples who greet the world anew.

Martial Culture

Sacred Groves

The sacred spaces of Scandinavian religion were associated with the forest, such as groves and trees dedicated to Thor. An example is the sacred grove of *Caill Tomair* near Dublin, which was destroyed after the defeat of Vikings by the Christian Irish forces led by Brian Bóruma in 999 AD (Colm). Neil Price, professor in the Department of Archaeology and Ancient History at Uppsala University, Sweden, characterizes the Norse peoples as the "children of Ash and Elm." Trees figure prominently in worship throughout Scandinavia and the veneration of ancestral trees is extant within the mythology.

Nature as a sacred place is a repetitive theme within various martial cultures. The concept of forests as a place of renewal and healing figures prominently in various individual and collective experiences for those within US martial culture. Several organizations host servicemembers, veterans, and their families in martial culture-only wilderness retreats of various kinds ("Wilderness Therapy for Veterans"). The experiences seek to use nature as a way of reconnecting and recentering the holistic self.

Those who understand that a martial lifeway requires progressing along a path of martial maturation often find natural world experiences a necessity. Freddy Osuna, a former Marine Corps Scout Sniper, now teaches wilderness tracking to all peoples and ages. His delight in traveling to different parts of the world and engaging with unique natural climates allows him to continue his relationship with nature. Jason France, a retired Air Force Chief Master Sergeant, took three months off immediately after his retirement to solo hike the entire Pacific Crest Trail. There is little doubt that the natural world is a place for spiritual rest today as it was for the Norse and their veneration of the protector deity Thor.

Valkyrie

The commonly understood conceptions of the *Valkyrie* are rooted in Sturluson's *Prose Edda* as well as texts such as Saxo's *History the Danes*. However, other texts demonstrate this is only one understanding of these supernatural creatures and that their names paint a very different picture of who the *Valkyrie* are. *Valkyrie*

> names evoke the trappings of war—weapons, noises, blood and corpses—suggesting nothing about their welcoming disposition that Snorri emphasizes [in the *Prose Edda*]. The skaldic representation of valkyries [… is that they] first and foremost […] choose the slain, and contrary to the impression given in Snorri's Edda, they are far from pleasant. (Fridriksdottir 5)

An exploration of *Valkyrie* names suggests they are named for objects or circumstances which can lead to death on the battlefield. For example, "Gondul, the 'War-Fetter', who brought the freezing hesitation that could be fatal" is a prime example of naming *valkyries* for an experience that occurs in battle (Price 54). Other names of *Valkyrie* include "Battle-Weaver, Shaker, Disorder, Scent-of-Battle, Victory-Froth, Vibration, Unstable, Treader, Swan White, Shield-Destroyer, Helper, Armour [sic], Devestate [sic], and Silence" (Price 54). The concept of *Valkyrie* as beautiful supernatural women seemed to evolve only after the Viking Age. Snorri's *Valkyrie* who ride above the waves and ships, choose mortal men as partners, and are present at the young god Baldr's funerary rites may be evolutions designed to make their roles more palatable in retellings.

While *valkryie* "are firmly supernatural" (Fridriksdottir 67), shield maidens were human women who appear in Icelandic sagas and Saxo's *History of the Danes* as "women warriors who receive training in battle skills and make a career of being Vikings" (Fridriksdottir 64). The separation in time between the women who appear in the sagas and those who recorded them challenges the notion that they ever existed. However, it is not conclusive that Norse women warriors did not exist.

Given the prominence of *valkyrie* and shield maidens in Norse myth and saga, women and their status among the Norse peoples must be examined before moving to any discussion of the mythologies and how they relate to martial culture. Norse warriors or Vikings are usually assumed to be male, but the various Scandinavian and Icelandic sagas—along with recent archeological discoveries—suggest otherwise.

Price argues there is a difficulty with "bias cascade" in terms of assumptions regarding gender in excavated finds in general, but of the Scandinavians in particular. Recent discoveries have thrown into question many conclusions drawn from burial sites which historically used ornamentation or weapons to classify the individual as female or male. One example is the warrior grave discovered in Birka, Sweden, in 1878, which was classified as male based on the accoutrements within the grave:

> The grave goods include a sword, an axe, a spear, armour-piercing arrows, a battle knife, two shields, and two horses, one mare and one stallion; thus, the complete equipment of a professional warrior. Furthermore, a full set of gaming pieces indicates knowledge of tactics and strategy. (Hedenstierna-Jonson et al. 854)

In 2011 "an osteological study suggested the buried person was actually female" (Price 177). The study "was considered controversial in a historical and archaeological context" (Hedenstierna-Jonson et al. 853) and therefore genetic testing was conducted. The genetic study concluded that "the individual in grave Bj 581 is the first confirmed female high-ranking Viking warrior" (Hedenstierna-Jonson et al. 857) and that the "results call for caution against generalizations regarding social orders in past societies" (Hedenstierna-Jonson et al. 853), while confirming some of the mythology and sagas regarding the roles of women in Scandinavian, and indeed, Viking society.

The women who suffered misidentification after death because they were buried with warrior accoutrements are followed today by women servicemembers and veterans who are not regarded as such because of their gender. Jennifer Blackmarr, a retired Air Force Senior Master Sergeant, remarks, "I hate getting the woman veteran questions too. Like my service was not the same as everyone else's service. Even at the VA" (Blackmarr). Martial cultures are not and thus should not be assumed to be completely dominated by male participants. References to "warrior" should be read as gender neutral, allowing for the reality that not everyone who wields an ax or sword is assumed to be male. Workers in a US Veterans Administration hospital should not always assume the woman who is in the halls is a spouse rather than a veteran.

Defender and Warfighter

Thor and Odin developed independently and gradually merged in the myths. "Odin is the supreme magician, master of runes, head of all divine society, patron of heroes, living or dead. Thor is the god of the hammer, enemy of the giants, whom he occasionally resembles in his fury" (Dumézil *Gods of the Ancient Northmen* 4). Popular understandings and retellings are derived from Sturluson's *Prose Edda* which relates that Thor is the son of Odin. Thor existed in Scandinavia while the Germanic *Wotan* eventually migrated out of mainland Europe and became Odin—a composite of *Wotan* and perhaps *Odir*²—after reaching Scandinavia. The tension between the two may stem from the potential conflict in traditions they represented.

> As early as the time of the Brothers Grimm Ludwig Uhland had pointed out that,
>   while Sweden worshipped Frey as its chief god, Norway paid homage to Thor. The god Odin he saw as a later importation from Saxaland to Scandinavia where he took hold mainly among the members of courtly society, owing chiefly to the proselytizing efforts of the court poets, the scalds [.... The] two Icelandic Eddas [were] a late and exclusively Scandinavian development. (Dumézil *Gods of the Ancient Northmen* xxii)

It can be surmised that the introduction of Wotan and his transformation into Odin came into conflict with the existing protectors such as Frey and Thor. Further evidence that the Germanic deity Odin conflicted with the indigenous Scandinavian deities may lay in the actual importance accorded Thor:

> "We hear more of the images of Thor than of those of the other gods, and when he shared a temple with other deities, he is usually said to have occupied the place of honour [sic] [...] His worshippers would look for guidance from the image of Thor when the time came to make some difficult decision." (Davidson *Gods and Myths of Northern Europe* 75)

For example, Adam of Bremen, a Christian cleric who had conducted missionary work in Northern Europe, made a record which included his journey. Bremen describes a great temple at Uppsala:

> In this temple, entirely decked out in gold, the people worship the statues of three gods in such wise that the mightiest of them, Thor, occupies a throne in the middle of the chamber; Wodan and Fricco have places on either side. The significance of these gods is as follows: Thor, they say, presides over the air, which governs the thunder and lightning, the winds and rains, fair weather and crops. The other, Wodan—that is, the Furious—carries on war and imparts to man strength against his enemies. The third is Fricco, who bestows peace and pleasure on mortals. His likeness, too, they fashion with an immense phallus. But Wodan they chisel armed, as our people are wont to represent Mars. Thor with his sceptre apparently resembles Jove. (Price 212)

The familial connection may be an attempt by Sturluson to create a Greek or Roman pantheon equivalent where none existed, as a way of simplifying and bringing Norse religion into the fold of more extant Mediterranean mythologies with which the Christians were familiar. Yet there is no argument that both were associated with warfare to a greater degree than other Norse deities, though all goddesses included, had an active role in combat.

While Odin is only concerned with the final war, Thor is the true guardian and friend to the Norse. Thor was considered a protector by the Scandinavians. As Davidson points out "In Dublin the Irish referred to the Viking settlers as 'people of Thor'" (Davidson *Gods and Myths of Northern Europe* 88–89). While the mythology depicts Thor as not only brutish but somewhat comical, the recorded stories do not seem to coincide with how the Norse actually viewed him. For while Thor guarded the realm of *Asgard*, he was "also regarded as the Defence [sic] of Men (aIda bergr)" (Davidson *Gods and Myths of Northern Europe* 91). Understanding the mythologies shifts dramatically in light of Davidson's assertions. Thor morphs from a sometimes comedic lout to one who is a fervent protector of humankind. As Davidson posits, Thor

> was struggling for mankind, and for the precarious civilization which men had wrested from a hard and chaotic world. [Thor] supported law, helped men to build and to cultivate, to marry and bring up children, and protected them on their journeyings. He guarded [...] the humbler homesteads of Norway and Iceland, marked out and hallowed by his sacred fire and his hammer-sign; he safeguarded their oaths with one another, and invested them with the sanctity of his temple and holy place. As the sky god, he drove his chariot over the circuit of the heavens, and could at will grant the traveler desirable weather and favourable winds. In Asgard he kept the goddesses of peace and plenty safe, so that they could grant their benefits to mankind; on earth, in the stony and storm-beaten lands of the north, he battled with the monsters of cold and violence that unceasingly threatened men's security. (*Gods and Myths of Northern Europe* 91)

Thor's immediate concerns are protection of the community and upholding of justice within social norms. The symbol of his hammer is used as blessing and protection of humans in all matters and can be found on numerous artifacts.

The reinvigoration of heathen religions is slowly being accepted within US martial culture, forcing the Department of Defense to officially recognize heathen religious practitioners. Additionally, the US Veterans Administration has authorized the symbol of *Mjölnir*—Thor's Hammer—on headstones of veterans who were followers of Asatru or Odinism (Horton). The use of the hammer symbol on a grave marker is a continuation of how the symbol was used by the Norse.

Thor's Hammer is "shown on a Swedish rune-stone from Stenqvista, where it is used as a sign of Thor's protection over the grave [... and is] found on stones bearing early runic inscriptions in Norway and Sweden, and some of these call on Thor to protect the memorial and place of burial" (Davidson *Gods and Myths of Northern Europe* 82–83). In this way, Thor's power is manifest in the symbol of the hammer and is associated with the health of the community. The hammer was present during "birth, marriage, and death, burial, and cremation ceremonies, weapons and feasting, travelling, land-taking, and the making of oaths" (Davidson *Gods and Myths of Northern Europe* 84). The placing of Thor's Hammer on a tombstone is symbolic of both life and death. *Mjölnir* was declared the finest weapon and dedicated from its creation to the defense of *Asgard*. It is illuminating that the chosen symbol for US martial culture adherents to heathenism, specifically Norse beliefs, represents the defender of the community, Thor rather than a symbol of Odin, the Norse god of war.

These definitive differences between the two are illuminated in the tension that exists between Thor's understanding of "keeping the peace" with Odin's more aggressive (and perhaps ruthless) focus on warfare by building his army through meddling in human conflict.

> Thor defends the ordered world (that is, the cosmos as habitable for humans and gods) against chaotic cosmic forces, whereas Odin engages in warfare amongst men, assisting those he has chosen to defeat others. Moreover, Odin has a strong motivation for his interference in earthly warfare [….,] gathering of a military army of the dead (einherjar) in Valholl, the hall of the valr. (Line 251)

Odin looks forward to *Ragnarök*, the great final battle that ends the cosmos, and is relentlessly obsessed with preventing the inevitable fall of the gods, as decreed in a prophecy related in the poem *Voluspa*. In this way, Odin is less concerned with earthly affairs than the final cosmic apocalypse of *Ragnarök* V. To that end, Odin represents those who seek to prevent future end-of-the-world calamity and as demonstrated in the various sagas, is willing to do whatever it takes—including the sacrifice of any individual, even himself—to deny the outcome of predestination found in the prophecy.

Thor, Odin, the Valkyrie, and others of the Norse pantheon represent various aspects of martial culture. But Odin and Thor are placed in contention with each other in the myths. Within the servicemember, in some instances, Odin's influence is too powerful and the servicemember or veteran is out of balance, so focused on the coming battle that he disregards all other things, rather than knowing that battles are as Thor understands them, an endemic lifeway to continuously work to not only protect the community, but serve it.

Perseverance

Thor's journey to Utgard-Loki's home and the ensuing tests are actually illusions for the true challenge. Because of his abilities, Thor has to be tricked into failing; yet his perception that he is failing made it real within his mind. Thor receives a gift of perceived failure. His ability to rally and demonstrate resilience against defeat after defeat reveals true emotional and spiritual strength. The Norse are also characterized as a people who greatly valued positivity and unrelenting perseverance in the face of fate. "A man who was prepared to die for what seemed to him important was held in honour [sic], whether friend or enemy, and won even greater admiration if he could die with a jest on his lips" (Davidson *Gods and Myths of Northern Europe* 10).

Resilient perseverance is a trait highly valued within the US military today. Failure tests character under extreme conditions and teaches far more than any success. Many servicemembers who go to advanced training will find that the true tests are the mental, emotional, and spiritual struggles which matter far more than physical tests. The limitations within each consciousness are so much more important to overcome. Jason France failed the US Air Force Pre-Ranger course—a prerequisite for Air Force personnel to attend the US Army Ranger Course—three times, finally passing on his fourth attempt, which opened the door to later graduating US Army Ranger School heralding a martial lifeway leading him to work with some of the greatest units in US martial culture (France).

Instructors can sometimes struggle to allow failures to occur. As Jennifer Blackmarr, a retired Air Force Senior Master Sergeant and former Basic Training Instructor notes, "It's hard for me to let them fail and I usually will see it happening even if they don't. I understand that they have to fail by themselves" (Blackmarr). The servicemember must fail, or be made to fail at some point, in order to learn how to conduct their day-to-day operations and cope with the unfairness of combat. In combat, a person can do everything correctly, yet that does not mean the outcome will be a success. The ability to bounce back means that a setback in combat does not lead to complete failure. This adage applies to all cultures, not just US martial culture. Resilience is the latest buzzword to be used within the US military organization to describe this type of mental, emotional, and spiritual recovery that is depicted in this mythical story of Thor's failure. The problem is not new, and so ancestral martial cultures included mythic examples of how to not be broken by failure, but to instead overcome doubt, and rally one more time to pick up the world serpent, as Thor did.

Suffering Gladly

Suffering for the sake of living fully and gaining knowledge or experience appears often in Norse mythology, specifically in the experiences of *Yggdrasil* and Odin, and especially the suffering they share when Odin seeks the secret of the runes.

The suffering of *Yggdrasil*, as described in the poem *Grimnismal* implies a sympathetic connection between the Scandinavian peoples and the suffering that the cosmic tree experiences in service of life. "The tree Yggdrasil/endures more pain/than any men guess./It's eaten from above by the deer,/on the side by rot,/from beneath by serpents" (Crawford 67). The description of the world tree that connects all realms through its roots and that *Yggdrasil* is constantly under attack—suffering for the sake of keeping the cosmos alive—demonstrates a little-known aspect of Scandinavian peoples in the way they relate to and fit in to the natural world.

Painful threshold experiences such as Odin's demonstrate the lengths one must endure to obtain wisdom.

> I know that I hung on a windswept tree/nine long nights,/wounded with a spear, dedicated to Odin,/myself to myself,/on that tree of which no man knows/from where its roots run./With no bread did they refresh me nor a drink from a horn,/downwards I peered;/I took up the runes, screaming I took them,/then I fell back from there. (Larrington 32)

Servicemembers volunteer to subject themselves to states of suffering in order to access special physical abilities, knowledge, and/or awareness. Freddy Osuna describes his decision to choose the US Marine Corps over the other services as an example.

> I didn't have any knowledge of what being a Marine was at all. I didn't know any Marines growing up. I didn't research it enough. I had a lot of preconceptions about what it was to be a Marine. I just knew I wanted to be one. And you know why? Because everybody I told I wanted to be a Marine in high school… they told me you don't want to do that. And these are grownups… these are adults telling me you don't want to do that Freddy. You don't want to do that…telling me to go in the Army or join the Air Force or the Navy. And that just made me want to go more! Right? Because I was just born to suffer. (Osuna)

The choice to submit to and even choose to sacrifice through physical and mental anguish is documented in the many documentaries or YouTube videos that show training of pilots, special operations personnel, and many other roles within the US military. Trainees experience moments in which anguish disappears and a type of euphoria settles, where one passes the threshold into a different state of existence.

The Ecstatic Berserk Gift

The berserker warriors of Odin, connected with the god through various texts, represent not only a devotion to battle but perhaps also fierce loyalty to a leader. They appear in the *Prose Edda* numerous times. Odin's berserkers are also present at the funerary rites described in *The Death of Baldr* and *Hermod's Ride to Hel*.

In the *Prose Edda*, King Adis, at war with King Ali, asks for King Hrolf Kraki to send his army to assist. Instead, Hrolf sends 12 berserkers, "Bodvar Bjarki was among them, and so also was Hjalti the Courageous, Hvitserk the Bold, Vott, Veseti and the brothers Svipdag and Beigud" (Sturluson 105). The story implies that because of the berserkers, King Adis carried the day. Snorri Sturluson also authored sagas in which the berserkers are described in greater detail. In the *Saga of the Ynglingas*, Odin's warriors

> went [into battle] without mail-coats and were as wild as dogs or wolves; they bit their shields, were as strong as bears or bulls, they killed people, but they themselves were hurt by neither fire nor iron; this is called going berserk. (qtd in Price 326)

King Hrolf Kraki and his berserkers are in lodgings provided by King Adil when Adil's men challenged Hrolf and his berserkers to withstand fire "so hot that the clothes were burned off King Hrolf and his men" (Sturluson 105). Hrolf and his berserkers then jumped across the conflagration.

Hrolf and his *berserkers* connect with deep and powerful energies that a berserker-as-shaman must harness and control. Odin's powers can cause "paralysis of the enemy troops, [and] 'madness' increasing by tenfold the normal powers of the favored soldiers" (Dumézil *Gods of the Northmen* 29). The berserker is also granted Odin's powers to access magical abilities, such as shape-changing and imperviousness to enemy weapons.

Popular contemporary understanding of *berserkers* as those who "go berserk" is inaccurate and does not represent the original berserker mindset. There is no easy or agreed upon definition of a Norse *berserker*. *Berserkers* are often described as special troops that either make up a specialized detachment within a larger force, are the personal guard of a leader, or both. Descriptions of varying reliability abound about both their wild nature and use of substances to heighten their fury. Most often the term *berserker* is translated as bear shirt, perhaps referring to bear skins the warriors might have worn, "That the bear connection may be the more relevant of the two is reinforced by a lupine counterpart to the berserkers in the form of the *ulfheðnar*, meaning 'wolfskins'" (Price 324). Yet *berserker* "itself refers to a shirt (*serk*) with either a *bear*- or *bare*- prefix," (Price 324). The later definition is perhaps partially informed by the descriptions of *berserker* actions in the *Prose Edda*.

The term berserker can refer to a state of being as in "*berserk*-fury" and a class of *berserker* is described in the poem *Haraldskvadi* that asks

> "what are they like, these men who go happy into battle? *Ulfheðnar* they are called, who bear bloody shields in the slaughter; they redden spears when they join the fighting; there they are arranged for their task; there I know that

the honourable [sic] prince places his trust only in brave men, who hack at shields." (qtd in Price 325; Kershaw 83–86)

Recall that *ulfheðnar* means wolfskins and in this passage it is the wolf, more than the bear, that is represented in *berserker* organizations. To further complicate matters, costuming—wearing bear or wolf skins—seems to have been a symbol of *berserker* warriors and there are many images of warriors with animal heads on pendants and textiles from the Viking period.

However, beyond the already varied potential meanings of berserker, Price offers yet another view of who the *berserkers* might have been and how their behavior may be interpreted:

There is no evidence whatsoever, in archaeology or text, for the berserkers' use of hallucinogens, entheogens, or any other form of mind-altering drug or chemical, including the consumption of fly agaric [...] The term *berserksgangr* (generally translated as 'going berserk', as above) literally describes a way of moving, 'berserk-walking', and not a fighting rage at all—something that might fit well with the strangely formal postures of the 'weapon dancers' and the pelt-wearing warriors on the metalwork. This could be a ritual, a sort of militarised [sic] performance, and one scholar has suggested that these theatrics were the real root of berserker-hood—effectively a symbolic preparation for war rather than reflecting any actual behaviour [sic] on the battlefield. (Price 326)

In short, to go "*berserk*" might be a type of war dance or performance. Neil Price and Kris Kershaw connect the *berserkers* to "'weapon dancer' images, men either naked or clad in wolf pelts and wielding swords and spears" (Price 324). In this light, *berserkers* could be viewed as a group of dedicated warriors who practiced a shamanic-warrior ability, not through hallucinogens, but in a manner perhaps similar to the whirling dervishes, i.e., ecstatic dance prior to battle to put themselves on a plane of existence that allowed for excellence and skill in warfare. This would not be dissimilar to the Māori Haka dance still in practice today in New Zealand military units.

It is clear from other sources, recorded from the opponents who encountered berserkers and *ulfheðnar* firsthand, that they were fierce warriors who howled like animals and fought with such fury as to be possibly described as resembling possession, a phenomenon likewise attributed to shamanistic peoples. Shape-changing and magic are often associated with shamans. Christianity demonized shamanic practices because it sought to supplant local religious and/or spiritual personalities, which may explain why the sagas, written from a Christian perspective, found useful fodder in judging *berserkers* as evil. Further, in Dumézil's latter group of Odinic heroes, Sigurd was his chief example of a warrior who demonstrated special abilities. Specifically, when Sigurd cooks the heart of the dragon and tastes the blood, he gains the ability to understand animal speech, a common trait among shamans.

Kris Kershaw, in his study on Odin, aligns *berserkers* with the concept of warrior-brahmins in the Hindu mythologies. The alignment of Norse deities and heroes with their counterparts from India follows Dumézil who postulated that there are Indo-European correlations in terms of religion and deities. Kershaw's correlation of berserkers and warrior-brahmin recalls the golden age which was ushered in by the union of the *kshatriya* and the brahmin, resulting in warrior-brahmin.

Ecstasy in combat or in shamanic ritual is often thought of in terms of a type of rabidness, a loss of higher cognition, and a settling into a baser, instinctual frenzy or incoherent trance. However,

> ecstasy does not mean loss of control. As Nora Chadwick writes in her study of shamans, 'it is important to emphasize the seer's perfect mastery and control, not only of his artistic material, but also of his own movements ....' [...] The divorce from reality is from the limitations of reality, allowing a total focus on the one skill or power, whether this is composing verses [...] or using weapons in a dance or a battle. (Kershaw 83)

Without discussing the idea of shape changing or *berserker* ritual war dancing, Jan Angstrom posits that there is a transformation that occurs when a modern Swedish servicemember moves from the garrison or base to the field. The servicemember experiences a "soldier to warrior transformation" (Haldén and Jackson 154) and becomes one with their natural environment by donning camouflage appropriate for the landscape.

The result "is a surprising ecological or environmental ideal of being one with nature in very tough terrain. Being invisible, operating behind enemy lines [.... In] Swedish, the English word 'ranger' means *skogsvaktare* [...] which literally means 'the one that guards the forest'" (Haldén and Jackson 155). Angstrom highlights that one soldier who was interviewed stated "in the field, I do not exist. It is only we that do" (Haldén and Jackson 154). Angstrom argues there is no sign of becoming "fierce beasts hiding in the woods [...but] independent-thinking individuals" (Haldén and Jackson 155). Both through study and experience, and in agreement with the soldier's quote above, the servicemembers become not independent, but interdependent, both on each other and the immediate surrounding world. Angstrom's argument highlights the transformational aspect of consciousness that must occur as one moves from every-day martial culture into the realm of becoming a part of an active operation. What is seen is not rabidness but control and mastery "allowing a total focus on the one skill or power" (Kershaw 83) which demonstrates the shamanic quality Kershaw argues for.

The views of Price, Kershaw, Dumézil, Angstrom, and Davidson combine to paint an elaborate multifaceted understanding of Odin and his *berserkers*.

As a type of ecstatic warrior group within the larger martial culture, the various descriptions and hypotheses of who they were give a fascinating account of warriors who may have used ritual dance and sound to attain and direct an ecstatic power that allowed them to shrewdly practice their art with an uncommon mastery.

Ecstatic warriors are common in martial cultures, yet the insistence on secular government service often fails to account for the ecstatic experiences of servicemembers. An industrialized or post-industrialized secularization does not make room for break-through ecstatic experiences that can occur as the result of the alignment of both internal and external factors that produce a power which is unbidden and for which the experiencer is potentially not prepared. Ritual preparation and guidance on how to enter into and then return from a different state of existence is not consciously understood by most servicemembers, though those who are aware occasionally develop unofficial models to meet this need without fully understanding the why of their necessity.

Martial cultures throughout the world have stories and examples of those who evidence every appearance of giving themselves over to a type of possession just prior to and during combat or in training for combat. The usual manifestations are described as akin to animal or natural phenomena of significant power. Examples include the embodiment of wolves, tigers, lions, and storms. The power of the natural forces of earth and sky is still in existence within US military units today in name, unit symbolism, and individual and collective experiences. There are servicemembers who experience possessive power during times of heightened intensive activity/situations in relation to a physical, mental, emotional, and spiritual single-pointed being who have no explanation for such an occurrence. Nor does the servicemember have anyone to turn to who can guide them through coming down from such an experience.

These altered states of consciousness need to be acknowledged. Further, as the attainment of these states can be entered into accidentally, servicemembers should be acquainted with these states prior to experiencing them on the battlefield so they can control the entering into and exiting out. Otherwise, without understanding or guidance, such experiences can create psychoses or problems adapting back to the typical martial cultural plane of existence. The ecstatic experience should not be feared and, while perhaps supernatural, is not unnatural. The warrior-shaman is not unusual, as witnessed in the warrior-brahmin discussed in the Hindu *Mahabharata*. The servicemember who leaves one plane of consciousness to attain another for the purposes of fulfilling their role within a martial culture is prevalent within US martial culture.

Ecstatic experiences of US servicemembers are plentiful, but rarely recorded. In some cases, the servicemembers lack the ability to describe the moment; for others, the experience is shared only within martial culture, perhaps because of the fear that anyone outside would condemn such experiences as a symptom of pathological issues and/or the subject as mentally defective. A former US Army

Ranger, Matt Best's brutal honesty, while abrasive to many, gives a rare glimpse into feelings that I have personally experienced and heard repeated in trusted circles:

> By the middle of this fifth deployment, I felt closer to that feeling than I ever had before. I never told anyone on my team about this [….] I didn't say it out loud, even as I was thinking it every second of every night we were out on target. The more intense a situation got, the deeper I went into the weirder parts of my brain. I would literally think to myself on missions at night, "[…] Good luck trying to kill me [….]" Don't misunderstand: I wasn't suicidal. Thinking you're going to die and wanting to die are totally different things. I didn't have a death wish. It's just that, in my experience, the more you deploy and face the dark realities that exist in life, the more comfortable you become with the idea of death. Sometimes you don't really care if it's you or the people you are hunting who die, just as long as it isn't the people you are leading. It's hard to explain to people who have never served in this capacity [….] I was peaking out at all levels, all deployment, in a way that I knew would stick with me wherever I went in life, however long that life might last. And the more I did it all, the more I wanted—the more I needed—to keep doing it. (Best et al. 120–22)

This kind of openness is rarely glimpsed outside of martial culture, and even then, it is only spoken to a trusted few. The need to continue to peak "out at all levels" is very common to those in martial culture, and not only in combat. Recall the amount of humanitarian and disaster relief work the US military is called on to do, the most recent being the deployment of medical personnel to supplement American hospitals during the Covid-19 crisis (Kimball). Peaking can be explained as elevating oneself to an existential plane that can only be described as a unification of body, mind, and soul linked to something beyond and yet so deeply embedded in the self that the experience is unable to be expressed within the limitations of language. Other veterans speak of how they had "'seen the bear and heard the hoot owl' together under fire" (Harris) or to put it simply, "We're […] animals" (Harris). Erik Elden, former member of the Norwegian Telemark Battalion, shares his reflections on being deployed:

> I experienced my second great change. A new consciousness on a spiritual plane, like the Indian [Native American] tests of manhood where you are exposed to fear, pain and loneliness. A situation where you feel that you are mortal, and are forced to accept it. To be alone in the wilderness, where survival is not guaranteed, but something to fight for. An experience with total focus in the moment and a constant acknowledgement of one's own vulnerability and mortality. (Elden as qtd in Dyvik)

This is a rarity to Best's and Elden's descriptive and raw accounts. To be so honest about the powers that course through one who is in touch with the shamanic nature of martial culture is to risk at best condemnation or at worst, be greeted by fear resulting in a psychological diagnosis and medication from well-meaning but uninitiated medical professionals.

The losing of oneself to action without being distracted or overcome by the chaos, to become the chaos without being subsumed by it, to control the tempo of the storm … these ideals should be the goal of every servicemember who serves in *extremis*. The need to access a different plane of being, to become something beyond what is typically considered "human," is something that many servicemembers strive to find in training so that they may feel comfortable when immersed in such power during moments of indescribable experience.

An example of this sort of extremity, which may seem incredulous but has many examples throughout US martial cultural experiences, is that of Roy Benavidez, a legend within US martial culture. On May 2, 1968, then-Staff Sergeant Benavidez heard cries for help over the radio from a Special Forces team consisting of US Green Berets and Montagnard tribesmen who had been ambushed by North Vietnamese forces. Without thought, Benavidez immediately boarded a helicopter flying out to rescue the team. He was armed only with a knife. Jumping from the helicopter, he was immediately "shot in the face, head and right leg" but continued forward, making contact with the team, four of whom were dead and the rest wounded (Goldstein). He got the survivors to the helicopter, but the pilot was killed as it took off, and the helicopter they were on crashed. Benavidez evacuated "the troops off the helicopter, and over the next six hours, he organized return fire, called in air strikes, administered morphine and recovered classified documents, although he got shot in the stomach and thigh and hit in the back by grenade fragments" (Goldstein). The fighting reduced itself to hand-to-hand combat as he continued to protect the team. At last, he was finally able to get the survivors onto a second helicopter and leave the field. "When he arrived at Loc Ninh, Sergeant Benavidez was unable to move or speak. Just as he was about to be placed into a body bag, he spit into a doctor's face to signal that he was still alive" (Goldstein). The close-knit collective martial culture demonstrates the connection an individual can have with another. In this unprepared eruption into action, an ecstasy occurs in which, like the berserker, neither fire nor weapons seem to have an effect on preventing a servicemember who is purposefully and precisely executing their role in combat for the communal aspect of their martial family. Ecstatic warriors, those that can call on Odinic powers to place them in a state of existence where they can leap through fire and where no weapons can touch them, exist within the myths of Scandinavians, yet these myths perhaps bear some relation to lived reality, as evidenced by the extraordinary feats accomplished by servicemembers.

US martial culture does not have an officially recognized place for the shamanic berserker states, either in preparation or in action. Many elements of the berserker state of consciousness are experienced as a breaking through during an action or battle. However, shamanic acceptance and preparation that trains servicemembers to experience and be comfortable at this level of consciousness or state of being would allow for a greater control and comfortable with the power rather than feeling the surge first in the heat of battle. Further, training would inculcate a methodology for exiting the state, descending back to a "normal" plane of existence without the journey without damaging the individual or the community.

Outcast

Recall the earlier reference to the *Gautreks* saga in which Odin and Thor bless and curse the Viking Starkad. Norse mythology provides abundant examples of the warrior as an outcast. Within the call and response between Odin and Thor, the concept of being a great warrior and yet living outside of society as an outcast is evident. While Odin grants the warrior "the best of weapons and clothing" (Kershaw 45) Thor retorts that Starkad "shall have neither land nor estates" (Kershaw 45) which in a society that prizes hearth and home acts as a curse, forcing the warrior to wander without a tether to a land or community. While Odin offers great wealth, Thor insists the warrior will "never be satisfied" (Kershaw 45), again implying an inability to be at rest within oneself. Odin grants victory in every conflict while Thor curses that each win will be at a grave cost. Odin grants the gift of composing beautiful poetry, yet Thor robs the warrior of his memory for Starkad "shall never remember afterwards" (Kershaw 45). The compositions are therefore locked within the warrior, robbing both him and the world of their beauty. And while the "noblest and the best" (Kershaw 45) will hold the warrior in great esteem, the "common people shall hate him every one" (Kershaw 45).

Many veterans and their families find themselves perceived as outcasts of both the military organization and civilian culture. Best describes his dissatisfaction with the prevailing narratives that reduce veterans to stereotypes:

> The one thing I kept coming back to—an issue that had become really frustrating to me—was the way people in our society talk about veterans. All you ever heard [...] were things like the destructiveness of PTSD [post-traumatic stress disorder] or the crippling nature of survivor's guilt. [...] Every veteran story was just this endless parade of horribles. What [popular culture and media] failed to show [...] was my experience, which was the same as the experience of the hundreds of veterans I've known and served with who loved their time in the military and to this day view it as one of the most important, meaningful, enjoyable periods of their lives. No matter where you looked, there was no appetite for our stories anywhere. It felt like the forces that controlled the culture, that attempted to shape how we reckon with war and

the warriors who fight it, had not built enough tolerance into the system [...] to accommodate the powerful notion that there are men and women out there who put their lives at risk to fight for others, to fight for an ideal, not because they had to but because they wanted to, they needed to. These were the forces that convinced civilians to thank us for our service on airport concourses all across America, in solemn, guilt-riddled tones, like we must have been compelled, reluctantly, to sacrifice our freedom, when in fact we had proactively exercised it to enlist and do something we loved. (Best et al. 186–87)

The idea of fully embracing martial culture, of volunteering to subject oneself to suffering, yet to do so gladly, and then ask for more, is very antithetical to most people's understanding of a desired lifeway. For that and many other reasons, not everyone who enters the military will successfully complete their term. And those who do may even stay for a long time but may not seek any challenges other than the minimum necessary to stay within the military organization. There are varying degrees of dedication to the welfare of the group over the individual in every collective culture. Yet there is a shared commonality of endured hardship filled with humor and love, sometimes that humor can be considered "dark" and as such colors popular understandings of those within martial culture and contributes to stereotyping of martial culture.

Many see servicemembers and veterans as purveyors of dark humor, seekers of violence, and largely indifferent to human suffering in war. This prevailing narrative overshadows the multiplicity of servicemember experiences that make up the entirety of a martial cultural self.

I actually don't tell most people that I was in the military for a couple reasons. One, they think women didn't do shit. So I had to hear of stupid things like that. And two, a lot of folks are already stereotyping us, or they think that we're crazy and hurt. (Blackmarr)

Vikings, as well as the wider Norse peoples, are often painted with a similar brush in popular narrative. Wider views such as those espoused by the archeologist Neil Price give a more holistic view of the Norse peoples. Similarly, more insider rather than outsider understandings of martial culture are necessary to break stereotypical treatment of servicemembers and veterans today.

Price emphasizes how fully the stereotype of those who lived in the Viking Age has been accepted and, using archeological records, demonstrates that the Scandinavian world of the Viking period "was a strongly multicultural and multi-ethnic place [....] and there is no doubt that a stroll through the market centres [sic] and trading places of the time would have been a vibrantly cosmopolitan experience" (Price 24). Yet Norse are considered barbarians in the same way civilian culture often stereotypes martial culture.

Unfortunately, many veterans or servicemembers resort to self-censoring whenever they find themselves outside of martial culture, avoiding even being identified as a veteran. As Best notes,

> my goal was to speak to people like me. People who [...] had no use for the pity; who did not need thanks for their service because they were more thankful for it than anyone could imagine. They were grateful for the chance to serve. I wanted to reflect their reality back to them so they would know that they weren't crazy for not being crazy. I also wanted any veterans and current active military who might be struggling to know that it was okay to laugh in the face of the horrors of war, that they could be proud of what they'd accomplished, and that there was at least one place online where no one would judge them either way. I wanted the world to know that veterans like me [...] weren't ticking time bombs waiting to explode. We were normal people who just so happened to have gone through some extraordinary experiences and come out the other side proud of our accomplishments, grateful for our brothers and sisters. (Best et al. 187–88)

The perception of being an outcast is strong (and perhaps, as a way to cope, cultivated and embraced) within veteran and their family communities. Most who are still in an active military organization are insulated from the extreme disparities in experiences, cultural norms, and daily habits that seem to appear between martial and civilian cultures when one is pitched out into the "real world." The tendency to become antisocial and avoid public locations may stem from one or two incidents that the veteran or their families experience when interacting with civilians that result in an embarrassment or negative judgment on either or both sides. These may be due to cultural incongruences.

Further, there is a sense that, once servicemembers leave the military organization, even if they continue in another capacity within government service as civilian employees working with the military, that they no longer "belong." They are the past and, in a very industrialized secular mentality, their experiences, knowledge, and advice are not always welcome and can even be met with hostility from the next generation who are currently serving. The idea that people are broken cogs removed from the industrial machine of military organization is one of the highest forms of soullessness that poisons martial culture. The veteran who experienced both these events is unwanted and unmissed by the military organization and shunned as pariah by civilian culture. Veterans can find themselves adrift in a liminal space that appears to lead nowhere.

While written in the late thirteenth century, the blessings and curses bestowed on Starkad show correlations in the experiences of US servicemembers today. Many veterans find themselves feeling cast out of their communities, perceiving that their experiences make them different. Yet they may also experience the phenomenon wherein, after a time, the people around them become frustrated and eventually begin to chafe at why the veteran cannot simply "get over" or dismiss existing as a veteran. This begs the question of why a veteran should choose to be ashamed of being a part of martial culture? While transformation is inevitable, there is no healthy way of moving forward other than by incorporating "veteranhood" into the future self. Suppression of that part of one's life (also referred to as self-oppression), hiding part of oneself because the wider society deems it unworthy, is damaging and antithetical to efforts today that seek to abolish oppression. The Norse peoples built metaphorical ways forward for veterans to live in peace within their beliefs. One is Odin's hall known as Valhalla. The other is Freya's hall *Folkvangar*, where the battle is done and living forward begins. To wish to become a guest of Odin's Hall is to want to continue to serve in an army; Freya's hall, *Folkvangar*, is the place for those who served well and are now ready to rest among others who served as they had. Perhaps contemplation about taking refuge in Freya's Hall will afford veterans a view future life as the gift it is, not a curse but still very much a part of the waft and weave of martial culture.

Renewal

Ragnarök is the final battle where the reigning generation of the gods of *Asgard* fall in a pyrrhic victory against Loki, the giants, and Loki's offspring. Each detail is foretold, including the rebirth of the world which heralds a golden age. But that golden age must be paid for by a great sacrifice, which is a common theme throughout both the mythology and the lives of the Scandinavians. The next world is one of peace and plenty under the divine guidance of the next generation of deities. They are led by Balder, the son of Odin who was "said to be the fairest of all and most deserving of praise; he was white of skin and bright-haired, and was both wise and merciful" (Davidson *Gods and Myths of Northern Europe* 29). Balder was killed very early in life by a projectile of mistletoe but is fated to rise and lead the new world. He is not alone, for "The sons of the great gods, like Vali, Vidar, and Magni, had special parts to play, for they were to inherit the world of Asgard when the older generation had perished" (Davidson *Gods and Myths of Northern Europe* 30). The need for the old generation to consume itself is necessary, despite the efforts of Odin to prevent it. Rebirth is promised in the *Voluspa*, and through that rebirth a world comes into being.

The *Vafthruthnismal* and *Voluspa* describe this new world. For example, to populate the world, this promise of rebirth is found in the *Voluspa* as well as other poems such as the *Vafthruthnismal*. The rebirth of the world and the continuation of the next generation after a great cosmic war is a motif found in many mythologies describing martial experiences.

There is a necessity to move on, past the deprivations of conflict, to find a life that uses the joy and pain of martial experiences as fertile soil to continue to grow in the light of a new sun. Perhaps veterans will revive the cry "To Folkvangar" rather than "To Valhalla," finding the spiritual and emotional peace of green fields and a great homestead in Asgard, which is similar to the Navajo Enemy Way Ceremony and the *Mahabharata*. The Navajo brothers, who, having fought their final battles and recovered from their illness, move forward in life. The Pandavas must move on to embrace and guide a peaceful new era. Renewal is a repetitive theme in many mythologies.

A golden age of renewal is the ideal, as mythologies are often wont to present. The reality can be very difficult to achieve, especially if one must exist in a different culture. In his book *Tribe*, Sebastian Junger asserts that people miss the experience of extreme situations such as war or natural disasters because those things bring people together in communal suffering and shared hardship. He argues that veterans miss war partly because "Modern society has perfected the art of making people not feel necessary" (Junger xxi). I would add the caveat that Junger is really describing an individualistic, post-modern secularism that emphasizes uniqueness to such an unhealthy extent that it encourages otherness. As discussed previously, service to a team/unit and others in need of protection is the bedrock of US military creeds. Inculcating an inappropriate degree of cynicism toward others or any type of cultural framework that attempts to create a sense of service creates a rot that eats away at the cords of shared humanity.

To this end, veterans have taken it upon themselves to establish their own sense of *communitas* through various veteran service organizations and informally to insure they stay in close connection with other servicemembers. The desire for these connections can make veterans more susceptible to organized criminal elements and extremists because they appear, on the surface, to offer familial kinship and a sense of duty similar to the military experience. Yet, a robust and heartfelt understanding of mythic martial heritage may serve to inculcate ways of perceiving true service from the hollow mirages offered them by these other groups. Groups who know quite well that many veterans may be under a new sun, but a harsh one in a trackless pitching sea that creates an unquenchable maddening thirst for the cool welcoming sight of land and the fresh water of tribal belonging.

For all who exist within US martial culture, whether they practice Norse religions or not, there are clear connections of shared experiences and commonality with the Scandinavian peoples who plied the oceans and waterways of Europe and preserved what they could in their mythologies. The wanderlust, the cosmopolitan nature of their lives, an interest in exploration and, yes, serving their communities in harsh climates while engaging in conflict, all demonstrate there is much to be learned from the texts that were written about their beliefs and myths. The most important may be the beautiful future that martial cultural adherents have within themselves to bring forth, just as the ashes of Ragnarök give way to an Earth that "will arise again from the waves, fertile, green, and fair as never before" (Davidson *Gods and Myths of Northern Europe* 38).

Notes

1 *Asgard* may have been envisioned by the feudal and agrocentric Norse as a land of great farmsteads, each of which centered around a magnificent hall.
2 A seldom mentioned god that appears in both the *Poetic* and *Prose Eddas*.

References

Best, Mat, et al. *Thank You for My Service*. Bantam Books, 2019.

Blackmarr, Jennifer. Personal Interview. 9 July 2023.

Crawford, Jackson. *The Poetic Edda: Stories of the Norse Gods and Heroes*. Hackett Publishing Company, Inc., 2015.

Colm. "Thor's Wood, a Sacred Grove Near Viking Age Dublin?" *Irish Archaeology*, 9 Feb. 2018, http://irisharchaeology.ie/2013/07/thors-wood-a-sacred-grove-in-viking-dublin/

Davidson, H. R. Ellis. *Gods and Myths of Northern Europe.* Penguin, 1990.

Dumézil, Georges. *Gods of the Ancient Northmen*. U of California P, 1977.

Dyvik, Synne L. "Valhalla Rising." *Security Dialogue*, vol. 47, no. 2, 2016, pp. 133–50.

France, Jason. Personal Interview. 28 June 2023.

Fridriksdottir, Johanna Katrin. *Valkyrie: The Women of the Viking World*. Bloomsbury Academic, 2021.

Goldstein, Richard. "Roy P. Benavidez, Recipient of Medal of Honor, Dies at 63." *The New York Times*, 4 Dec. 1998, https://www.nytimes.com/1998/12/04/us/roy-p-benavidez-recipient-of-medal-of-honor-dies-at-63.html

Haldén, Peter, and Peter Jackson, editors. *Transforming Warriors the Ritual Organization of Military Force*. Routledge, 2016.

Harris, Art. "The Great Green Beret War-Story Weekend." *The Washington Post*, WP Company, 8 July 1980, https://www.washingtonpost.com/archive/lifestyle/1980/07/08/the-great-green-beret-war-story-weekend/b8ed3966-3987-4e21-b0e6-e2f835523e3d/

Hedenstierna-Jonson, et al. "A female Viking Sarrior Confirmed by Genomics." *American Journal of Physical Anthropology*. vol. 164, 2017, pp. 853–60, https://doi.org/10.1002/ajpa.23308

Horton, Alex. "The Hammer of Thor: Now Approved for VA Provided Headstones." *VAntage Point*, Veteran's Administration, 13 Nov. 2013, https://blogs.va.gov/VAntage/9906/the-hammer-of-thor-now-approved-for-va-provided-headstones/

Junger, Sebastian. *Tribe: On Homecoming & Belonging*. Fourth Estate Ltd, 2017.

Kershaw, Kris. "The One-Eyed God Odin and the (Indo)-Germanic Mannerbunde." *Journal of Indo-European Studies Monograph No. 36*, 2000.

Kimball, Spencer. "U.S. Will Deploy Troops to Help Hospitals during Omicron Wave, Distribute Free COVID Tests Starting in January." *CNBC*, 21 Dec. 2021, https://www.cnbc.com/2021/12/21/omicron-us-to-deploy-troops-to-hospitals-purchase-500-million-covid-tests.html

Larrington, Carolyne. *The Poetic Edda* (Oxford World's Classics), Kindle ed., OUP Oxford, 2014, p. 32.

Line, Philip. *The Vikings and Their Enemies: Warfare in Northern Europe 750–1100.* Skyhorse Publishing, 2015.

Margaryan, A., Lawson, D. J., Sikora, M., et al. "Population Genomics of the Viking World." *Nature*, vol. 585, 2020, pp. 390–96, https://doi.org/10.1038/s41586-020-2688-8

Osuna, Freddy. Personal Interview. 14 June 2023.

Price, Neil. *Children of Ash and Elm: A History of the Vikings.* Basic Books, 2020.

Sturluson, Snorri. *The Prose Edda: Norse Mythology.* Translated by Jesse L. Byock, Penguin, 2005.

6 Japanese

This chapter examines the myths and religions of Japan that formed the basis of ancient Japanese warrior cultures and explores how these myths resonate with US martial culture today. Combined with the challenging geography of Japan and the idea that "Japanese myths present a somewhat disorganized pattern, [and are] episodic rather than epic in nature" (Saunders qtd in Kitagawa 31), exploration of this mythological tradition and martial culture is organized by theme rather than a chronological exploration of a single text.

There were many martial cultures within ancient Japan, and many martial lifeway experiences were recorded by the warriors who belonged to them. To support the exploration of these themes, it is necessary to briefly explore the roots of the Japanese people and sources of the energy of Japanese myth and martial service.

The Japanese word used to describe the lands and waters of their world is *Nihon* or *Nippon,* commonly interpreted as the "sun's origin" or the "Land of the Rising Sun." The peoples of these islands embraced the sun, as demonstrated by the Rising Sun flags of Japan. Those flags, known as *kyokujitsu-ki,* have a red sun disc on a white background with 8 or 16 red rays emanating from the sun disc, respectively. The original flag was flown by various feudal lords (*daimyo*) during the Edo Period (1603–868) and was adopted as the de facto national flag in 1870 at the beginning of the new Meiji State.

The sun disc symbol makes an appearance in the *Tale of the Heike* (*Heike Monogatari*), a dramatized history of the tumultuous events of twelfth-century CE Japan. This epic combines historical events with supernatural influences and includes a riveting scene of a band of samurai gathered beside a river at sunset who see "a stunning girl, not yet twenty, wearing a fivefold willow layering and a red hakama divided skirt, came to the side [of the river] and planted there, upright, a red fan bearing a sun disk" (Tyler 594). This description of a sun disc symbol on a fan (banner) borne by a young woman seems to encapsulate, in the dying light of a setting sun, an incarnation of the goddess *Ama-terasu-oho-mi-kami,* who is usually referred to more simply as Amaterasu.

DOI: 10.4324/9781032613222-6

Thus, the Japanese flag invokes the sun goddess Amaterasu, inviting her presence into all things. In this guise, she has been present in Japanese martial culture from the earliest mythologies, which are related in two texts: "*The Records of Ancient Matters (Kojiki)* and the *Chronicles of Japan (Nihon-shoki* or *Nihongi)*" (Kitagawa 142). The *Kojiki* is considered the primary mythological text while the *Nihongi*, incorporating elements of the *Kojiki*, is considered a more historical text. The *Kojiki* tells of the initial creation myth of Japan.

Mythic Origins

To understand the importance of Sun Goddess Amaterasu, one must understand the origin mythology of the Japanese people and the importance of the underlying principles regarding the term *kami*, which traces its origins to the mythic birth of the world. She is kami, which is

> an honorific for noble, sacred spirits, which implies a sense of adoration for their virtues and authority [....] Among the objects or phenomena designated from ancient times as kami are the qualities of growth, fertility, and production; natural phenomena, such as wind and thunder; natural objects, such as the sun, mountains, rivers, trees and rocks; some animals; and ancestral spirits. In the last-named category are the spirits of the Imperial ancestors, the ancestors of noble families, and in a sense all ancestral spirits. Also regarded as kami are the guardian spirits of the land, occupations, and skills; the spirits of national heroes, men of outstanding deeds or virtues, and those who have contributed to civilization, culture, and human welfare; those who have died for the state or the community; and the pitiable dead." (Ono and Woodard 6–7)

All creatures and places may be considered potential *kami*. In this way, sacredness is everywhere.

While both the *Kojiki* and the *Nihon-shoki* (or *Nihongi*) describe the events, Kitagawa asserts that "the *Kojiki* shows far less Chinese influence [...] and for that reason we will follow [it] for the discussion of cosmogonic, and other myths" (142). Basil Chamberlain notes in the introduction to his 1882 translation of the *Kojiki* that it "is the most important [text] because it has preserved for us more faithfully than any other book the mythology, the manners, the language, and the traditional history of Ancient Japan" (i). Therefore, the *Kojiki*, first recorded in text by Ō no Yasumaro in 712 CE, is the closest to the textual record of the origin myth of Japan.

Whereas the beliefs of the earliest peoples living in Japan are largely lost, Joseph Kitagawa argues that their cosmology survives, with Chinese influence, in the Japanese origin myths *Kojiki* and *Nihongi* (Kitagawa 18–19, 31). The roots of these myths are fed by the sources of the energies of Japanese myth. *Kami* are the gravitational centers of the Indigenous belief system. Amaterasu is a *kami,* as are all of the ancient deities who created the world. Additionally, past emperors, teachers, and samurai are also *kami* (Mason and Caiger 32). *Kami* have an underlying multiplicity which contributes to intense ambiguity. Kitagawa notes the word *kami* "is [also] used both to designate an impersonal quality, [...] the kami nature, somewhat analogous to the numinous or sacred, and [...] for specific beings endowed with the kami nature, be they human, divine, or other animate or inanimate beings" (Ono and Woodard 36). It should be noted that to imbue *kami* attributes only implies supernatural power, not a morality. "Eminence here does not refer merely to the superiority of nobility, goodness or meritorious deeds. Evil and mysterious things, if they are extraordinary and dreadful, are [also] called kami" (Norinaga qtd in Mason and Caiger 33).

The roots of Japanese myth grow from Shintō, literally, the "Kami Way." The word Shintō "is composed of two ideographs 神 (shin), which is equated with the indigenous term *kami,* and 道 (dō or tō), which is equated with the term *michi,* meaning 'way'" (Ono and Woodard 2). It is also referred to as *Kami No Michi* (the way of the Kami).

Shintō was a late invention intended to delineate the Indigenous Japanese beliefs from the imported belief systems such as Buddhism[1] and Confucianism.[2] The disparate beliefs that form Shintō have the collective understanding that Shintō is a kami-faith, implying belief

> in the kami, usages practiced in accordance with the mind of the kami, and spiritual life attained through the worship of and in communion with the kami. To those who worship kami, "Shintō" is a collective noun denoting all faiths. It is an all-inclusive term embracing the various faiths which are comprehended in the kami-idea [....] Shintō is more than a religious faith. It is an amalgam of attitudes, ideas, and ways of doing things [....] Thus, Shintō is both a personal faith in the kami and a communal way of life according to the mind of the kami." (Ono and Woodard 3)

Shintō shrines dedicated to *kami* were usually established in places where encounters with the numinous occurred, but also in villages, and today "there are tens of thousands of shrines throughout Japan" (Mason and Caiger 34). Shintō is the basis for Japanese beliefs, and while "foreign influences are evident [....and the] kami-faith cannot be fully understood without some reference to them [Sokyo Ono argues] it is as Indigenous as the people that brought the Japanese nation into existence" (Ono and Woodard 1). The lands of Japan are alive with *kami* power and their individual selves are reflected within their natural wildness.

The isolating geographical nature of Japan may have been a contributing factor to the multitude of mythic "natures" found throughout the lands and seas. For the "islands, by their very shape, ensured that communications would remain a problem [....with] 80 [sic] per cent of the land surface [...] mountainous" with rich vegetation and "many a swift flowing river and clear lake, and [...] a climate that gives sharply defined seasons of warm humid summers and snowy winters, dramatic springs and melancholy autumns" Turnbull asserts that in early Japan, "merely to keep alive must have been an adventure" (*The Samurai* 3).

The sheer number of Shintō shrines acknowledges that *kami* are ubiquitous and as "for the individual shrines, there was no connection whatsoever between them, not even among those in a given geographical area or those that had the same objects of worship" (Ono and Woodard 17). *Torii* mark the entry to *kami* shrines, and are "the gateway which symbolically marks off the mundane world from the world of the kami, the secular from the spiritual" (Ono and Woodard 28). Because each island was born of a *kami* the entirety of Japan is considered sacred space, and by extension, the entire cosmos.

Shintō demonstrates that nature and human are one. Hayao Kawai asserts that in Japan "the concept of Nature is quite different from that of the West" (25). In medieval Japanese dream stories, "there was no distinct demarcation between life and death, reality and fantasy, myself and others. The same holds true for man and Nature" (Kawai 26). The Japanese term *shizen*, often translated as Nature, "expresses a state in which everything flows spontaneously. There is something like an ever-changing flow in which everything—sky, earth, and man—is contained" (26–27). Kawai claims that to understand the connection, one must embrace both *shizen* and *jinen*, which is "more comprehensive than 'Nature'" and encompasses *shizen*. Therefore, to understand the relationship between Japanese and the world is to accept that you "and I, humans and Nature, reality and fantasy, flow spontaneously in *jinen*, which transcends all distinctions" (27). *Jinen* is at the root of Japanese myth and religion and is integral to an understanding of how Japanese warriors understood their role in life and death. With this understanding of the relationship between *kami,* Shintō, and how *jinen* (Nature) encompasses all things within a sacred environment, it is possible to discuss the origin story that establishes the importance of *kami* in the heart of Japanese myth.

The Kojiki

To understand the mythological time of Japan, it is necessary to understand that mythology

in early Japan [...] are not childish imagination or superstition [....] these myths preserve the early Japanese understanding of the meaning of the world and of life, and more especially the mode of existence of kami. To them, all the events mentioned in myths—the marriage between two kami or their method of planting seeds and weaving—took place in mythical time. [....] As such, myths provided a heavenly model for earthly life, and in this sense "religion" embraced all aspects of existence of the people of Japan during the prehistoric and early historic period. (Kitagawa 31)

In the beginning, there only exists a primordial ether and then earth and the High Heavens manifest into existence. In the High Heavens three *kami* came into being. This spontaneous eruption of separate life is "defined by Motowori as 'the birth of that which did not exist before'" (Chamberlain and Aston 17). These three *kami* self-manifested separately and then hid themselves. In the same way, two more *kami* came into being and also concealed themselves. After these first five *kami*, the next seven generations of *kami* are considered "the first seven generations of spirits" (Yasumaro 8).

The first five generations of *kami* willed themselves into being and then hid themselves.[3] The idea that the very first *kami* are present but hidden insinuates a kind of spirituality wherein the universe has layers of hidden sentient sacredness woven into its very fabric. The Japanese seem to regard the next six generations of *kami* as connected to the seventh generation, who created Japan. The seventh generation consist of a brother and sister, Izanagi (He Who Beckoned) and Izanami (She Who Beckoned). They are commanded to create land and were given a "jeweled halberd of heaven" which they dipped into the sea and began stirring, churning the sea. As they removed the halberd from the waters, great gatherings of salt dripped off the tip forming an island. The creation of the first primal earth isle happens through the joined actions of female and male *kami* using a bequeathed spear from the *kami* host of heaven to stir the waters. Then, the two descend to the island and begin coupling, creating 14 islands (Yasumaro 13).

After the islands are born, Izanagi and Izanami create 35 more kami. However, in birthing Swift Burning Flame Man, Izanami is severely wounded and becomes ill, creating six more children from her various bodily reactions as she succumbs to her wounds. Izanagi, moved to intense grieving, "unsheathed the sword ten hand spans long that was girded by his mighty side and beheaded his child," Swift Burning Man (Yasumaro 13).

Devastated, Izanagi seeks his sister in the underworld. When he finally finds her after journeying deep into the earth, he beholds her rotting body. Izanami is furious at Izanagi's seeing of her current form. He incurs Izanami's wrath for looking on her as she now is. Izanagi flees—pursued by his sister, who now commands a legion of underworld warrior *kami*—and fights a desperate battle with his sword to keep the kami warriors at bay. He is able to seal his sister and her army in the earth by blocking the entrance. Utterly exhausted, he rests beside a river. Realizing he is polluted by his visitation to the underworld, he needs to purify himself by kneeling at a river and bathing. Izanagi drops his staff and as he removes his garments, each item he becomes a spirit (12 in all). He bathes in different waters, creating 14 spirits throughout the process. Of the last three he proclaims, "After making child after child, I have at last gained three noble children!" (Yasumaro 18). In this act, 14 *kami* are born; however, the greatest of these are the three born from the cleansing of his face:

> Now he washed his mighty left eye, and the spirit named the great and mighty spirit Heaven Shining [Amaterasu] came into being. Next he washed his mighty right eye, and the spirit named the mighty one Moon Counting [Tsuku-yomi] came into being. Next he washed his mighty nose, and the spirit named the mighty one Reckless Rushing Raging Man [Susa-no-o] came into being." (Yasumaro 18)

Amaterasu, as first daughter, is given her place in the sky as Heaven Shining to rule over the day, Tsuku-yomi, as first son, is given place as Moon Counting (or Moon Bow) to rule over the night. Susa-no-o, the youngest son, is given place to rule over the seas; however, he rejects his father's charge, leading to further adventures.

Amaterasu is both a radiant goddess who gives life-light to the world as well as a fierce fighter when challenged. When Susa-no-o leaves the earth to approach her in her station in Heaven, Amaterasu suspects him of trying to seize her lands and

> she straightaway undid her mighty hair and swiftly parted it into two mighty looped locks to her left and right [; over] her back she slung a quiver filled with a thousand arrows; by her side she strapped on a quiver filled with five hundred arrows. Her bow arm was guarded by a stern bracer of bamboo. Brandishing aloft the upper end of her bow, she stomped hard on the firm-packed ground of the courtyard, kicking clods of earth up to her thighs [...] as though they were a light flurry of snow. Stamping her feet fiercely, she waited for her brother." (Yasumaro 19–20)

Amaterasu attempts to make peace with her wild and insubordinate brother, but they clash. In one confrontation, she witnesses her brother Susa-no-o wreak havoc upon the land and shuts herself away in "Heaven's Boulder Cavern" (Yasumaro 23) leaving the world in perpetual night. The self-imposed exile results in the world beginning to suffer from lack of sunlight. All of the *kami* collect at the entrance, entreating Amaterasu to come out. All attempts fail. Therefore, a mirror and a jewel were created to aid in bringing Amaterasu out from the cave in which she had taken refuge. The kami create a great commotion, and when Amaterasu looks out to see what is going on, a mirror is held aloft by two of the *kami*. Amaterasu, intrigued, steps out, whereupon she is pulled completely out of the cave by several *kami* to ensure she cannot retreat, thereupon restoring the world with her life-giving light.

Eventually she makes peace with her brother, striking a final balance of cooperation. Susa-no-o, striking out into the world, ends up battling and defeating a great dragon. He then proceeds to break his sword, cutting the middle tail.

> "Then, thinking it strange, he thrust into and split [the flesh] with the point of his august sword and looked, and there was a sharp great sword [within]. So he took this great sword, and, thinking it a strange thing, he respectfully informed the Heaven-Shining-Great-August-Deity." (Chamberlain and Aston 75)

This sword, Susa-no-o, presents to Amaterasu. It is named Grass Scyther or Grass Mower (*Kusa-nagi* or *Ame-no-Murakumo-no-Tsurugi*).

Amaterasu's mythological grandson, Ruling Rice Ears of Heaven (*Ninigi-no-Mikoto*), descends to earth, charged with ruling the divine earth of Japan. His great grandson, Emperor Jimmu, descendent of the Sun Goddess, eventually inherits three items: the sword *Kusa-nagi*, the mirror *Yata no Kagami*, and the jewel *Yasakani no Magatama*. "Some understand the mirror, jewel, and sword to symbolize the virtues of wisdom, benevolence, and courage respectively" (Ono and Woodard 25). The jewel is thought to complement the sword. "The sword and jewels, which are enclosed in brocade and are hung from the banner standards [...] symbolize both the power to defend the kami from evil and the power of the kami to protect justice and peace" (Ono and Woodard 25). The sword and jewel are kept together, amplifying the *kami*-nature of their protection. These are the symbols of the divine Emperor and exist to protect Japan.

Martial Culture

Creation and Death

> In this creation myth we find the first statement of certain aspects of Japanese tradition, [...] the divine ancestry of the Japanese rulers, and from the appearance, right from the beginning of time, of the symbol, of a weapon." (Turnbull 1)

Izanami and Izanagi create the islands of Japan through using a spear to stir the waters illustrates the foundational nature of martial culture in relation to the Japanese origin mythology. The obverse of life giving is life taking, and in the *Kojiki* the first death, specifically of the spirit, comes during the process of creation.

Izanami's death following her birth of Swift Burning Man, and his subsequent death by Izanagi, which in turn creates life, demonstrates that both creation and destruction are intimately linked within the myth. Death lives with those in martial culture, whether they are front-line combat personnel or support. While the goal in martial culture is one of service, that service is unique in the accepted understanding that this lifeway includes the use of force, up to and including deadly force. Many servicemembers are aware of the connection, yet it can be easily forgotten when one moves from training to daily activities that consist of cubicle office work rather than the dirt of the field.

Servicemembers train to cause death, prevent death, and work to rescue others from death. Physical training seeks to strengthen readiness to do all of these things. The advanced first-aid training of today is a gift from those who died from the lack trained medics nearby during World War II, Korea, and Vietnam. Basic marksmanship training for combat support and support personnel is designed more to protect oneself than to attack another, just as disparate villages in early Japan had to provide for their own security. Navy basic training requires trainees to drop into water from a ten foot platform jump, and perform both a 50 meter swim and a five-minute float to simulate jumping off a sinking ship and surviving in the water awaiting rescue. Officers in training maneuver symbols on maps, representing units of friendly, enemy, and civilians, and plan which targets are to be attacked. Death is a constant companion in the most mundane tasks, whether consciously acknowledged or not. Death is not necessarily framed as something to fear, but to acknowledge as a part of martial lifeway. The *Kojiki* describes the initial death and how the great *kami* who created the islands also experienced death.

The Japanese samurai Yamamato Tsunetomo wrote in the *Hagakure-kikigaki*—possibly the greatest work on *bushido* (often interpreted as the way of the warrior)—that true *bushido* "is to be found in dying" (Yamamato and Bennett 11, 42). Dan Oberg has written of how the *samurai* utilized the concept of death to transform into their warrior selves by placing the contemplation and acceptance of death at the center of their existence. He proposes death is approached in the *Hagakure* through three interpretations. One "illustrates the way in which warrior transformation takes place through the act of death or killing, or through meditating upon this act" (Haldén and Jackson 122).

Thus far in this book, the concept of training for, visualizing, and then actualizing intentional injury or death to another person has been largely glossed over. Whether an individual clears an aircraft for takeoff or is the pilot, or serves food to those who go out on patrol, or operates a turret-mounted machine gun, all have made a conscious choice to participate in or support the action of deliberate killing. Martial cultures seek to cultivate training in which one must visualize each action in advance. Practice requires repetitive visualization of future action with complete awareness that your life as well as the life of your team or community is dependent upon not failing at the critical moment. Whether a bow is flexed with an arrow notched or a firearm loaded, the practitioner must visualize all aspects of their actions toward a successful outcome.

One practice is to train in mock battles against one's peers. In combat exercises, there is a simulation of killing and being killed, in unarmed combat training where a servicemember uses rubber knives or utilizing training cartridges (blanks) in the servicemember's weapon. When "hit," the person falls down and lies on the ground to simulate being shot. Medics are called to treat the fallen person and therefore must visualize the type of simulated wounds and how to treat them.

So the first thing that really happens is the servicemember goes through a series of visualized killing, dying, and resurrection experiences in a cycle that simulates multiple lifetimes. This happens perhaps hundreds of times in a world of training and visualization in which one kills and is killed and is resurrected. At first, in many cases, the gravity of the gravity of the acts in which one goes through a constant series of rebirths is largely glossed over, approached as a game, and really not given the full *gravitas* merited. However, once one has actually gone to war, the reality of what "dying" in training can mean becomes apparent and the full measure of the necessity of training breaks through.

The samurai concept that to follow this lifeway means to accept death, both of yourself and of others, is similar to the daily life in US martial culture. Suzuki's assertion that a samurai must expose "himself before the enemy's swordstroke" (Suzuki and Jaffe 89) to fulfill a warrior's duty, and Tsunetomo's assertion that the "way of the warrior (*bushido*) is to be found in dying" (Yamamato 42) sum up how death is the central pivot point for a warrior. In US martial culture, death is ever-present, yet in practice it seems very much ignored, unless servicemembers understand fully the meanings behind the things they do in training. The looming aspect of death is a consistent state of being within US martial culture.

Kawai tells the story of Myoe (1173–232) who was born into a warrior family and became a Buddhist priest (Kawai 41). Myoe, at the age of 13, decides "I have become old enough" (Kawai 45) and goes to a place where human bodies are laid out on the ground, exposed to become food for local dogs and wolves. Myoe lays down that night awaiting death. "Wolves eventually appear, but only eat dead bodies, and not Myoe" (Kawai 46). Kawai goes on:

One of the reasons for Myoe's brave but naïve intention might stem from his warrior heritage. When he fails to carry out his self sacrifice, he realizes that he cannot die if fate does not decree it even though he himself has decided on it. [....] I imagine Myoe feels this kind of completeness, and has the notion that this completeness would be destroyed if he kept his life. (Kawai 46)

Kawai asserts that the familial warrior influence provides the impetus for acceptance to decide to embrace death. In a moment of sharing his story, Freddy Osuna speaks about his realization of accepting the idea of personal death.

You know, I found myself paralyzed in fear sometimes. This was after I lost a couple of guys down south by Najafi and I was terrified. But then one day I'm like, I'm tired living like this. I got to be more effective than this as a leader. [....] And so when I accepted it, I was okay. I mean, it wasn't like alright, I'm going to die. [....] It was…this is my calling… this is what I'm supposed to do. If I die today, that's okay. I'm fighting for my country and what I wanted to do. I put myself here. I accepted it. After that I was more effective. So I think that there is a certain point in major combat operations where you have to accept your own destiny. And, and after that point, there is a lot less to fear in this world. (Osuna)

Osuna's experience of overcoming fear of death in extreme circumstances is echoed in the story of Myoe. Myoe lets go of a fear of death by intentionally placing himself in mortal danger, coming through the trial unharmed and with a new consciousness that is freed from an affliction of the fear of his mortality. As Osuna illuminates in describing his epiphany about his own fear, once this fear has been overcome, a servicemember can blossom in abilities and potentialities. However, there is a path where one decides it is not enough to "accept your own destiny" but to fulfill it through one's own hand.

The Japanese samurai practice of ritual *seppuku* as a means of honorable death runs contrary to today's views of suicide. "*Seppuku* (i.e., ritual suicide by disembowelment, vulgarized in the West as hara-kiri)" (Fusé 57) is the ritual conducted by samurai to end their life. In some cases, like the case of Saigou detailed below, a second individual assists by beheading the intended, if the individual is unable to complete the disembowelment.

In 1877, Saigou Takamori led the final gasp of official samurai life in the form of a rebellion against the secular, industrial, and modernized military of Japan.

The imperial army began its final assault attack [on September 24] at 3:55 A.M. The [samurai] rebels defended their hilltop positions but were rapidly beaten back by superior force. By 5:30 A.M. the imperial army had destroyed the rebels' fortifications. The army moved artillery into these positions and began to concentrate fire on the valley below. Saigou's force was reduced to about forty men. At roughly 7:00 A.M. Saigou and his troops descended the hill to face the Japanese army and die. Saigou was surrounded by his closest and dearest allies: Kirino Toshiaki, Murata Shinpachi, Katsura Hisatake, and Beppu Shinsuke. Halfway down the hill, Saigou was shot in the right hip. The bullet passed through his body and exited at his left femur. Saigou fell to the ground. According to legend, Saigou composed himself and prepared for seppuku, samurai ritual suicide. Turning to Beppu he said, "My dear Shinsuke, I think this place will do. Please be my second (*kaishaku*)." Saigou then calmly faced east, toward the imperial palace, and bent his head. Beppu quickly severed his head with a single, clean stroke […]. (Ravina 4)

This assisted suicide, far from being condemned as a sin, is considered admirable, for the "ritualized death of a fallen hero was complete. Saigou had died a model samurai. *Nishikie*, colorful woodblock prints […] expanded on this legend [….] Saigou was shown, glorious and noble, pushing a sword into his abdomen" (Ravina 5). Saigou, who traced his lineage to one of the oldest samurai families, found defenders and advocates among the diversity of Japanese. While "the intelligentsia defended Saigou in essays, the populace defended him through legend and rumor" (Ravina 7) going so far as to elevate him to the level of a Buddha and claiming he had taken his position in the heavens as the bright planet Mars (Ravina 7–9). The Japanese government was forced to pardon and elevate Saigou to imperial hero, emphasizing samurai virtue, the very thing political power had sought to wipe out (Ravina 11–12).

Another "form of seppuku [is] called *junshi* or 'suicide to follow one's lord to the grave.' [….] Probably the best known *Junshi* in Tokugawa Japan was the group *seppuku* of the famous 47 *rōnin* (masterless samurai)" (Fusé 59). *Jigai*, another form of ritualized suicide, does not require seconds:

[T]he *jigai* ritual is a traditional method of female suicide, carried out by cutting the jugular vein using a knife called a tanto. The *jigai* ritual is the feminine counterpart of *seppuku* (well-known as *harakiri*), the ritual suicide of samurai warriors, which was carried out by a deep slash into the abdomen. In contrast to *seppuku*, *jigai* can be performed without assistance, which was fundamental for *seppuku*. (Maiese et al. 8)

Mark Blum asserts that *jigai* is linked to Buddhist concepts of relinquishing the body by "'achieving the Pure Land through suicide'—the most common motive given for Buddhist ascetic suicide in Japan" (Stone and Walter 10–11). Through the abandonment of the body in service and/or devotion to another or an ideal, the concept of taking one's own life as an accepted practice is extremely antithetical to today's understanding of lifeways.

In recent years, roughly 22 servicemembers and veterans commit suicide each day. This rate is widely considered an epidemic in terms of scale of death. According to the Veteran's Affairs

The number (count) of suicides among U.S. adults increased from 29,580 in 2001 to 45,861 in 2019 (see Figure 1). Veterans accounted for 5,989 suicides in 2001, which represented 20.2% of suicides among U.S. adults in 2001; and 6,261 suicides in 2019, which, by comparison, represented 13.7% of suicides among U.S. adults in 2019. Veterans ages 55–74 were the largest population subgroup; they accounted for 38.6% of Veteran suicide deaths in 2019. ("2021 National Veteran Suicide Prevention Annual Report.")

Suicide, in the Western mind, is a tragedy that affects all who know the individual, depriving those who love the individual of their active, physical presence, and is viewed as a loss of a beautiful potentiality of future life. The goal of US veteran health care and various veteran organizations is to eradicate this practice through prevention and treatment of the conditions that lead a person to choose to take their own life. In some cases, the smaller goals are set to restrict veterans and servicemembers from easy access to the implements that make suicide attempts more successful.

However, the Japanese viewed the taking of one's life as a way to demonstrate love and loyalty. As with the samurai who used their own sacred blades to deal with the killing stroke, the analogue might be modern firearms servicemembers carry.

Servicemembers are aware of, and live with, death as a present reality. Perhaps the trend to hide death away from western populations by using hospitals, caregiving facilities, coroners, and other services designed to shield the people from dead bodies, even of loved ones, exacerbates the problem of accepting death as a part of life. I have no answer; however, the exploration of the Japanese martial view of life opens the conversation and challenges Western paradigms and potential stigmas associated with the relationship between life and death.

Sword

"We do not have to read far in the ancient chronicles before we find mention of the first actual sword in mythology" (Turnbull *The Samurai* 4). Izanagi wields Sweeping Blade of Heaven. With it, he kills Swift Burning Man. In so doing, the blood that drips off of the great sword creates 16 more kami, all "born from the sword" of Izanagi. An instrument that both takes and creates sacred life, the sword is holy in Japan and inextricably linked to the samurai.

Samurai are described in detail in the *Tale of Heiki* in which two samurai clans, the Taira and Minamoto, usually referred to as Genji and Heiki, are the primary combatants in the Gempei War that lasted from 1180–85 (Turnbull *The Samurai* 40). Samurai "comes from the verb 'samurau' or 'saburau' which means 'to serve'" (Turnbull *The Samurai* 18) and originally had no military connotation. The samurai arose as small landowners and supporters of larger more powerful landowners who organized in the outlying provinces to defend themselves from attacks by groups such as the *emishi*, who were "skilled archers and fierce fighters" (Turnbull *The Samurai* 17). Eventually, the overarching landowners would take on the term *daimyo* and intermarriage among the samurai families would lead to allegiances of clans united under a *daimyo*.

Samurai were not all equal and "in many ways gradings within the samurai class had a greater rigidity than the division between that class as a whole and the rest of the population, since daimyo and other high-ranking military families formed an exclusive group, as did court nobles and outcasts" (Mason and Caiger 221). There were ten samurai ranks and they varied widely. "Men in the two lowest ranks—*yoriki* and *ashigaru*—were generally restricted to menial posts such as guard duty," therefore if one was a member of the eighth rank they would be considered "near the bottom of 'white collar' urban samurai" (Ravina 24).

Even then, highly ranked samurai families could find themselves living in poverty and potential homelessness, as Mark Ravina demonstrates in his biography of Saigo Takamori, a samurai who led a rebellion against the modernization of Japan and died in 1877. Caste standing was not a way of measuring affluence within Japanese society and most samurai held "day jobs" such as farming or fishing to earn enough to live on, not unlike the US National Guard and Reserve military servicemembers of today.

The term bushi is connected with samurai "and may be regarded as a general term for a fighting man or warrior" and is often found in the combination of *bushi-do* usually translated as "the Way of the Warrior" (Turnbull *The Samurai* 18). Thomas Cleary notes the

> "culture of the samurai, incorporating elements of Taoism, Confucianism, Buddhism [Zen], and Shintō, along with the martial arts and military science essential to their original profession, came to be referred to as budo, the warrior's way; or shido, the knight's way; and finally bushido, the way of the warrior-knight." (Cleary 1)

Bushido consists of numerous writings by various samurai, forming the philosophical and ethical principles of how to live a martial lifeway, along with practical advice in the arts of war.

One focus of samurai was their relationship with their weapons, particularly the sword. Suzuki describes this connection.

> "The sword is the soul of the samurai": [....] The samurai who wishes to be faithful to his vocation will have first of all to ask himself the question: How shall I transcend birth and death so that I can be ready at any moment to give up my life if necessary for my Lord? This means exposing himself before the enemy's swordstroke or directing his own sword toward himself. The sword thus becomes most intimately connected with the life of the samurai, and it has become the symbol of loyalty and self-sacrifice. [....] The sword comes to be identified with the annihilation of things that lie in the way of peace, justice, progress, and humanity. It stands for all that is desirable for the spiritual welfare of the world at large. It is now the embodiment of life and not of death. (Suzuki and Jaffe 89)

The *samurai* were distinguished by carrying the *katana* (long) and *wakizashi* (short) swords in public and these weapons came to mark their status to a point; to carry these two swords was illegal for anyone who was not of the samurai class.

The concept of their primary weapon as an embodiment of a soul reinforces that *kami* are within all things. It is not the sword but the blade that truly signifies the soul of the samurai, for the sword handle and fittings were and are viewed as merely dressing to represent the state of the blade which, in turn, signified the state of the samurai. *Koshirae* swords have a blade, handguard, fittings, and a handle, usually with cord wrapping. Swords that are dressed in this way are expected to be used often. When they are not, the handguard, fittings, and handle are removed and replaced with plain wood handle and scabbard and the sword assumes the designation *shirasaya*. *Shirasaya* swords, when sheathed, create an airtight seal and the appearance of a single piece of wood, all to protect the blade within. The mounting of the sword is a reflection, not just of the sword, but of the owner of the sword as active within the martial world sphere or in a retired state, their soul within strong and sharp but hidden. The concept of the "blade as soul" has a great correlation with modern struggles as servicemembers become veterans.

Purification

In the Japanese view of death, those who encounter it must be purified before returning to the "living." This is documented in Chinese writings: "The passage that follows is taken from a third-century A.D. Chinese history that included a section on the land of Wa (Japan) [....] When the funeral is over, all members of the whole family go into the water to cleanse themselves in a bath of purification" (Mason and Caiger 19). Purification, appearing extremely early in Japanese funeral rites, is found in the *Kojiki* in relation to the death of Izanami and Izanagi's flight from the underworld to the surface.

Izanagi is pursued by Izanami and the forces within the Underworld. He is forced to fight a rearguard action against

> "the eight thunder spirits [who] were also sent from the land of the Underworld in pursuit of him, accompanied by a force of fifteen hundred warriors. And so he unsheathed the sword ten hand spans long that was girded by his mighty side and waved it behind him as he fled. Still they pursued him." (15)

Izanagi fights a terrible battle. Rearguard actions are the last unit in a retreat that can protect the rest of the retreating forces from a pursuing enemy and are desperate battles, as experienced by US servicemembers during the Korean War in the Battle of Chosin Reservoir. The Chinese attempted to encircle and destroy the US Marines, soldiers, and Republic of Korea soldiers that made up X Corps. Those who survived the initial attack had to take a

> "narrow route southeast to the evacuation ports, surrounded by high ground that favored the pursuing Chinese, was characterized by incessant and fierce fighting along its flanks as Marine Corps, Army, and ROK units retreated, interspersed with hundreds of civilian refugees." ("Chosin Reservoir")

Donald Mason, a Marine Corps veteran who fought at Chosin, returned to Korea for ceremonies commemorating the sixty-ninth anniversary of the battle and described that time as "hell," perhaps analogous to the underworld Izanagi is attempting to flee from (Keeler). Once the forces and refugees reached the port of Hungnam to be evacuated by the Navy, their exhaustion, relief, and sadness and their need to, as Izanagi does, sink to their knees and cleanse themselves is understood. Those servicemembers might have well agreed with Izanagi's words: "How foul it is, this foul and filthy land I have been to! I should bathe to cleanse my mighty body" (Yasumaro 16). The gauntlet traversed by Izanagi bears a great resemblance to the experiences of those who fought through the mountains of the Korean peninsula.

Purification is a consistent theme in martial cultures. A sacred act combined with time and space for renewal is necessary. To be in contact with the otherworld is to bring some of it back into the world a servicemember is trying to protect. Therefore, methods of purifying are necessary. Recalling the stories of Mary Hegar and her impulse to swim in the Indian Ocean and Jason France's need to traverse the Pacific Coast Trail, it must be recognized that purification is still something that must be connected to *jinen*, for servicemembers are, as has been argued here, of "the wilds." Rest may be found in connections with the natural world.

Warrior Goddess

Born from her father's left eye, Amaterasu exhibits a steadfast warrior nature. She is also the greatest of Izanagi's children. Women samurai like Amaterasu are known as the *Onna-Bugeisha* or *Onna-musha* and served in Japanese martial culture. Stephen Turnbull asserts that over

> a period of eight centuries female samurai warriors are to be found on battlefields, warships and the walls of defended castles. Their family backgrounds range across all social classes from noblewomen to peasant farmers. Some were motivated by religious belief, others by politics, but all fought beside their menfolk with a determination and bravery [...] and, when the ultimate sacrifice was called for, they went willingly to their deaths as bravely as any male samurai. (Turnbull and Rava 4)

The martial feminine is at the root of Japanese myth and is ever-present in the flags of the modern Japanese military.

This book has already discussed the concept of gender in relation to martial cultures as well as that of the US martial culture, and the welcome changes that have begun to appear in attitudes and treatment of women servicemembers. The *Onna-Bugeisha* in Japanese martial cultures of the past as well as the mythic Amaterasu demonstrate the recurring theme of women as servicemembers.

Mirror

In the *Jinnō Shōtōki*, Chikafusa Kitabatake speaks of Amaterasu's mirror:

> the mirror hides nothing. It shines without a selfish mind. Everything good and bad, right and wrong, is reflected without fail. The mirror is the source of honesty because it has the virtue of responding according to the shape of objects. It points out the fairness and impartiality of the divine will. (Ono and Woodard 23)

The concept of looking at oneself and accepting all that the mirror impartially shows is difficult, because the reflection shows only what currently is, not necessarily the sum totality of the self, in terms of reflecting all the self has been, or will be. In "ancient society [the mirror] was an object of ceremonial and religious significance [....] Ninigi-no-mikoto was told by the Sun Goddess 'to honor and worship the mirror' as 'her spirit.'" (Ono and Woodard 23).

Mirrors are often attributed with the power to show truths perhaps hidden from the rest of the world. The Veteran Art Project, started by photographer Devin Mitchell, uses trick photography to create moving art that uses mirrors to illuminate the soul. Mitchell, utilizing a room, mirror, a veteran subject, and Photoshop to create a juxtaposition by taking two photos. The first is "a picture of the subject in uniform, the other in civilian attire" (Gibbons-Neff). Then Mitchell combines the two with the resulting image showing "a man staring into his bathroom mirror and adjusting his suit. Staring back is the same man, Lt. Ricky Ryba, in blue Navy fatigues. The resulting image transcends time and place" (Gibbons-Neff). This art project has attracted many requests from veterans who have found the experience cathartic as a single image encapsulating the liminality of what Mitchell calls the "double-life" (Gibbons-Neff) veterans wake to each day as the Sun Goddess Amaterasu rises into the sky.

For some servicemembers, identity becomes confusing when they exit the military organization. Many people look to personas, masks to be donned and doffed as needed for different situations. One may be a Sergeant at work and father and spouse at home, a child on the holidays when visiting parents. However, martial culture is an immersive world and as such, touches each part of a person's life. The Veteran Art Project depicts what many veterans and servicemembers feel regarding their deepest identity. As Brady McCoy, a retired Air Force Chief Master Sergeant described how he associates the Native American Chiefs with his own Chief stripes. And that he "learned how to be a good parent" from being a servicemember (McCoy). Once in martial culture, a veteran's identity will be tied to that lifeway, however brief the experience may be.

Hachiman, Jizo, and Buddhist Warriors

Several mythical pantheons have a mythological deity (or deities) who is given a permanent place as deity of war or conflict. It might be assumed that there is no such deity in Japanese mythology, given how interwoven myth is with Buddhism, which is famed for its position on nonviolence. However, the *Kojiki* demonstrates that conflict is practiced when necessary, by nearly every deity who is actively engaged with the earth. Yet these deities are not considered symbols of war. Further, a uniquely Japanese approach to all religions practiced by Japanese is the core of the preservation of Japan:

> Shintō, as Japan's first and basic religion, had the emperor as its high priest; [… and] his first concern and main function were the preservation of his people in safety and prosperity. Therefore, in times of crisis, such as the Mongolian invasions in the late thirteenth century, Shintō priests assiduously prayed to the gods and believed that they had responded by sending the gales (*kami kaze*) that had wrecked the Mongol fleets [.… Buddhism was adopted] primarily as a more potent means to the same end—preservation of the nation [.…] When the Mongols attacked, Buddhist clergy joined their sutra chanting to the ritual efforts of the Shintō priests to bring victory. (King 32–33)

Shintō and Buddhism amalgamated "Hachiman, the Shintō god of war, into a bodhisattva of high rank, [and] Buddhist warriors could pray to him for victory as well as Shintōists could" (King 32–33). The bodhisattva, Usa Hachiman, "with ancient roots in northern Kyushu, won a position of prominence at the Yamato court in the eighth and ninth centuries and then rose to new heights in the medieval period as the protecting deity of the Minamoto lineage that founded the shogunate" and had several shrines and temples across the country, with the central "Hachiman shrine-temple complex called Tsurugaoka Hachimanguji" serving as the "Kamakura's ritual center" (Breen and Teeuwen 42). This bodhisattva was originally a Shintō "deity of obscure origin" (Kitagawa 248) but has been linked to the "spirit of the ancient Emperor Ojin" (Breen and Teeuwen 78–79), who appears as one of the imperial descendants of Amaterasu in both the *Kojiki* and the *Nihongi*.

Hachiman appears in *The Tale of Heiki* amidst the prayers of samurai: "With one hand Yoichi covered his eyes and silently prayed the following prayer: 'Hail Hachiman, Great Bodhisattva, and you, gods of my home province, Nikkō Gongen of Utsunomiya, Yuzen Daimyōjin of Nasu, I beg of you, guide my arrow'" (594 Tyler). The earliest official mention of Hachiman is in the *Shoku Nihongi* and Ross Bender asserts that recent research has done much to explore the ancient beginnings of the Shintō Hachiman. It is enough to understand that "it is not generally believed that Ojin was associated with the Hachiman cult from its very beginning" but that by the time the Buddhists had consecrated Hachiman as a bodhisattva, the Ojin connection had been established (Bender 127–29).

The explorations of Hachiman have, however, demonstrated that characterizing Hachiman as a god of war may be, just as asserted earlier in this work regarding Mars, a vast oversimplification:

> Japanese scholars, however, have looked beyond this medieval phase of the Hachiman cult and begun the task of interpreting earlier stages of the belief. A number of Western writers have also discovered the complexity of the faith. U. A. Casal, despite the title of his article (*'Hachiman, der Kriegsgott Japans'*), concluded after a brief historical sketch and examination of the popular cult that the war-god aspect was only part of the picture—Hachiman was to a great extent a protector and preserver of life in non-military contexts. (Casal found, incidentally, that Hachiman's powers are so broad that women bring infants to the shrine to pray for protection against intestinal worms.) [....] The resulting picture of Hachiman is one of an amorphous deity with myriad functions. (Bender 127)

Hachiman may be described as less a god of war than a god of warriors. He is prayed to for strength and hope in times of crisis. Through the idea of "temporary transformation (*gonge* or *gongen* in Japanese) of buddhas and bodhisattvas," (Sadakata and Nakamura 134) later Buddhist Zen "samurai could [...] invoke the aid of the gods (kami) to assist him, especially that of Hachiman, the Shintō god of war, as well as a Buddhist bodhisattva, without betraying his Buddhism [by committing violence in combat]" (King 177).

Another bodhisattva shares traits with Hachiman in the way of a "protector and preserver of life" and also has an intimate connection with the protection of infants, women, children, and travelers as well as "liminal times and places, with marginal people, and with the special dead" (Glassman 8): Jizo. Jizo is associated with in-between spaces and is "a bodhisattva characterized by motion" (Glassman 11). As a psychopomp, he is the protector of spirits occupying the hells of Buddhist cosmos and his "association with the underworld, judgement in the next life, and [rescuer] from hell" (Glassman 15) asserts his primacy in that time of crossover from life to death. "In the sutras describing Jizo's past lives and his present mission, he is the savior of sentient beings in the period of 'a world without a Buddha'" (Glassman 7). Jizo's importance to all peoples who are in need is evident, yet there is also a particular connection between Jizo and the warrior.

While there are several manifestations of Jizo, one is called Warrior Jizo. Jizo is associated with the Pure Land *Amida bodhisattva* and various samurai families "championed the Jizo cult" through the efforts of Ritsu, Zen, and Pure Land monks (Glassman 76). As Glassman asserts, much effort was expended to align the warrior class with Jizo (76). Jizo is also connected to sacred dance for the

> "Wakamiya—especially its hall for sacred dances, held and extremely important place in religious life [.... The] principle image, *honzen*, of this hall was a Jizo statue. In fact, through the medieval period in various locales, religious performance was most commonly held in or near Jizo's halls." (99)

These halls, where dance, plays, and rituals occurred, were patronized by the *samurai* class. Glassman laments that the relationship between Jizo and warriors [...] deserves much fuller treatment [....] From Jizo's iconographical echoes in the sogyo Hachiman (the "monk form" of the Minamoto clan god, Hachiman), to the importance of the shogun Jizo in the founding legends of Kiyomizudera, to his cult at Mount Atago, to personal devotion by generations of samurai leaders [...] the "warrior Jizo" is a major theme awaiting investigation." (212)

The connection between warrior Jizo and Hachiman was so strong that in "1202, Chogen instructed Kaikei to make a Hachiman image 'in the form of Jizo' for Todaiji" (Hirabayashi as referenced in Glassman 213).

The connection and need for the warrior aspect of Jizo is rooted in Jizo's nature as a savior in the liminal and damned places where no other bodhisattva can be found. The connection between Jizo and Hachiman seems fitting because the warrior lifeway was made hereditary with rise of the samurai class and the emphasis on contemplation of death as the ever-present reality in samurai clans.

The presence of Hachiman and especially Jizo, who, like the Roman Mars has many aspects, is demonstrative the sacred and compassionate aspects of being a servicemember. This theme is recurring, as the unity of the holy and the warrior, healer and protector, represent the multipolarity of those who serve in martial culture.

Zen

Perhaps the most famous connection between myth/religion and martial culture in Japan is Zen Buddhism. Zen philosophy stems from Indian Buddhism which uses meditation as a way of overcoming the emptiness found in binary thought. Zen thought asserts that "enlightenment was not intellectual but existential and experiential" (King 10). Indian Buddhism was introduced to China and integrated with the "wildness and whimsical quality of Taoism" (King 14). While Confucianism had introduced the *Tao*, Buddhism introduced the *dharma* to the Japanese (Kitagawa 45, 51). The Taoist elements that Zen incorporated "were never fully house-broken [by the traditional Buddhist methods of retreat from the world, scripture study, and ritual-laden ceremony] and Zen glories in the fact that they were not" (King 14). This combination of Indian Buddhism and Chinese Tao became Zen. According to Daisetz T. Suzuki

Zen is discipline in enlightenment. Enlightenment means emancipation. And emancipation is no less than freedom [.... and] real freedom is the outcome of enlightenment. When a man realizes this, in whatever situation he may find himself he is always free in his inner life, for that pursues its own line of action. Zen is the religion of *jiyū*, (*tzŭ-yu*), "self-reliance," and *jizai (tzŭtsai*), "self-being." (Suzuki and Jaffe 5–6)

Eisai, a Tendai monk ordained as a Zen Master, is often regarded as the founder of Zen Buddhism in Japan. Yet it was Hojo Tokiyori who made Zen the "unofficial-official religion" (King 29) of the warrior-rulers of Japan.

> In Japan, Zen was intimately related from the beginning of its history to the life of the samurai. [....] Zen has sustained them in two ways, morally and philosophically. Morally, because Zen is a religion which teaches us not to look backward once the course is decided upon; philosophically, because it treats life and death indifferently [.... Zen is] a religion of the will. (Suzuki and Jaffe 61)

Zen contains an air of stoicism mixed with recognition and reverence for the experiential nature of communing with the sacred. Such a moment may be described as a peak experience, in which one grasps the nature of life while witnessing beauty in a blossom, with full acknowledgement that it will die. And at the same time, Zen embraces the transcendent that shines through the illusion of life and death.

Hayao Kawai, in describing key differences between Japanese versus Western thought, asserts that

> "Japanese values are based on the principle of completeness rather than of perfection [....] it is said that Japanese like imperfect beauty, or that they think the state of imperfection is more beautiful than the state of perfection. I think it is much better to say that there is a tendency in the Japanese to appreciate the beauty of completeness." (Dreams 119)

Kawai's assertion seems to be heavily informed by Zen. Kawai relates a

> famous story about a Zen master who shows what beauty is for him. A young monk is sweeping a garden. He tries to do his best at the job. He cleans the garden perfectly so that no dust is left in it. Contrary to his expectation the old master is not happy about his work. The young monk thinks for a while and shakes a tree so that several dead leaves fall down here and there in the garden. The master smiles when he sees that. (120)

Within the US martial culture, there is the concept of a prevailing need to ensure no mistakes are made, ultimately striving for perfection. Yet without those mistakes, one is not complete and cannot learn and grow. The space required to come back after failing makes for a more holistic person within martial culture. The mistakes and growth from missteps completes the character of the servicemember.

Another view of perfection versus completeness or wholeness in relation to US martial culture is how, when a servicemember becomes a veteran, they see themselves in relation to their actions as servicemembers. Many times, servicemembers simply do not experience or receive the sense of completeness or wholeness when reflecting on their actions during their time as servicemembers.

A tension may exist if the servicemember/veteran has the motivation to go into martial culture to accomplish something that makes a difference. Something palatable, something that the veteran can say "yes, I was one of them and we did something that mattered." Instead, US servicemembers have experienced ongoing rotations of units in and out of combat zones, being relieved after 4, 6, or even 18 months. Servicemembers have returned after a period of time to the same location, only to find that, from their perspective, nothing has improved. This experience quickly destroys the illusion of having done something positive and inculcates a cynicism that spreads like a psychological toxin throughout units and extends a sense of futility that endures throughout the servicemember's life.

To complete a mission or a task closes the circle and inculcates a sense of completion. Select units within the US military are given the latitude to receive a mission, plan all aspects, conduct the mission, and recover from it. For those fortunate few, there is a sense of completion not felt by others. Many servicemembers have no sense of the complex reasons to accomplish the mission. They are simply cogs in a larger machine. Once their part in the mission is complete, those servicemembers are not exactly sure what was supposed to happen; from their viewpoint, nothing happened correctly, even though three levels above them, that experience bore out the intended goal. Therefore, the perception is that neither perfection nor completion is achieved, which can lead to a sense of emptiness. A focus on completion, even if imperfect, may serve to allow for healing and moving forward in martial culture with satisfaction. Kawai's view of completion in the Japanese mind as of greater importance offers insight into the folly of relentless pursuit of perfection in the US military while failing to address the necessity of completion.

Reverence for Ancestors

As the lineage of the Imperial line goes back to the mythological age and the kami from which the royal family is descended, so too does the example of reverence for those who came before. Both the importance of, and respect for, familial elders, as well as reverence for deceased ancestors, is demonstrated throughout Japanese myth. The deification of passed ancestors allows for their transformation into *kami*, thereby keeping the ancestors both alive and aware of their descendants' actions and needs. These *kami* become guides and protectors of the family and their descendants. In *The Tale of the Heiki*, a petition is sent to a monastery to "invoke the divine aid" (Tyler 361) of *kami*. In the petition, there is a declaration of intent to rely on ancestral deeds and "righteous warriors" (Tyler 361) in their efforts to restore the kingdom. Reverence for those who have gone before abounds in Japanese mythology. In US martial culture, this practice has been maintained to a far lesser degree, but does exist.

US martial culture has embraced a type of organizational and cultural structure derived from an industrial model. Efficiency is prized over wisdom, and retirees and veterans are usually summarily cut off from those still active within the military organization. Only in historical study (of which there is quite little) does the current martial culture come near to approaching the type of reverence found in the ancient Japanese martial mythologies. Because of the secular nature of the US military, many servicemembers are only moved to study that history when it is required by the military organization for advancement in position or rank.

While many US servicemembers and veterans are not connected through blood ancestry, as the spiritual inheritors of martial culture, many find a growing kinship connection with servicemembers who have gone before them, and in a sense, have become a kind of sacred kami for US servicemembers and veterans. When US Marine recruits go to sleep at night, the last thing they do is bid a hearty good night to Chesty Puller. They are bringing to consciousness Lieutenant General "Chesty" Puller who served 37 years in the Marine Corps, in both officer and enlisted positions. Puller was a veteran of the "Banana Wars" in Haiti and Nicaragua, World War II, and the Korean War. The recruits who not yet Marines are being reminded of a deceased spiritual ancestor, verbally engaging "Chesty" as if he is in the room attending them as they move from consciousness to unconsciousness, potentially entering a dream-state, or perhaps a liminal lucid dream-state in which they are unconsciously meditating on the sacred example of a Marine.

Some military leaders keep a painting titled "The Prayer at Valley Forge" by Arnold Friberg in their offices or homes, as a type of meditative shrine. The scene depicts a profile view of George Washington kneeling in the snow next to his horse, head bent low, hands clasped. The painting captures one of the lowest points a leader in martial culture can experience. Washington had just suffered defeat and withdrawal from New York and moved what was left of the Continental Army into Valley Forge for the winter, where his military and their supplies dwindled. He lost many of his soldiers, not just to death, but to desertion as well. Using this scene in a type of meditative contemplation practice pulls the individual into a place where one comes to a quiet point, creating an introspection that allows for a ritualistic examination of current state of being.

More formalized ritual memorials mark remarkable events, such as the annual remembrance of those that perished during the attempted Iran hostage rescue mission in April of 1980, more well known as Operation Eagle Claw or referred to simply as Desert One. This ritual takes place at Hurlburt Field, Florida home of the US Air Force Special Operations. The unit that lost airmen that day was the 8th Special Operations Squadron. During the ritual remembrance, current servicemembers of this unit call the names of those lost.

One by one, Airmen stepped forward and took their place beside the podium, representing [the deceased] Capt. Richard Bakke, Capt. Harold Lewis, Capt.

Lyn McIntosh, Capt. James McMillan II, Tech. Sgt. Joel Mayo, Marine Staff Sgt. Dewey Johnson, Marine Sgt. John Harvey and Marine Cpl. George Holmes Jr. (Gonzales)

An elder, retired Colonel Kenneth Poole, a navigator that was at Desert One that day, addresses those present. He speaks to "the Airmen gathered that they had a responsibility to carry on the Desert One legacy. 'All special operators carry the flame to never fail our nation again,' said the colonel. 'All Air Commandos carry that flame today.'" (Gonzales) Then he solemnly salutes those lost that day. "'The hairs on the back of my neck stood up,' said Airman first Class Harry Tabata [....] 'It made me remember we were celebrating the birth of Air Force Special Operations Command and remembering our fallen comrades'" (Gonzales). There is a spiritual inheritance that occurs as the airmen step forward to "become" the fallen ancestral warriors at Desert One as Poole, the "gray-haired" elder, speaks the story of their ancestors forth. The combined effect of the ritual momentarily allows participants such as Tabata to connect to the transcendent. The deification of ancestors enables their transformation into a form of *kami*. These *kami* imbue natural power within the concept of *jinen* (only in the West would these be seen as supernatural) and become protectors of those human descendants or the lands they once protected in life.

Jonathan Ebel's work *G.I. Messiahs* discusses how American Battle Monuments Commission cemeteries such as Normandy, Flanders Field, and Manila consecrate and make the fallen US servicemembers sacred. This deification utilizes burial patterns whereby war "heroes lie buried next to cooks, nurses, and those who never got close to the battlefield. The dead are arranged without regard to class, race, creed, or national origin" (Ebel 97). While Ebel argues that the cemeteries are used as a form of civil religiosity to propagandize American democracy, from a martial culture lens these exist, much like the Shinto shrines, for servicemembers, veterans, and their families.

While the deification of ancestors in Japan most likely made *kami* out of people who might not have lived up to the warrior ideal, the deification that matches the *kami* model is far more selective than that outwardly displayed in US military cemeteries and their purpose as Ebel argues for. To be deified, a person must be of great renown. Within US martial culture, there are past ancestors who are very likely thought of in the same was as *kami*.

Identifying those who are truly deified within US martial culture begins with the examples provided above and continues by looking for those who, having passed this life, are remembered by those in martial culture through ceremony, ritual, and whispered by name before sleep, as if a prayer to one's goddess or god. That goddess or god, however, is usually one that is believed to have the most influence over the lifeway an individual chooses. In many cases, martial cultures of the past have had a specific deity of war or conflict.

Rōnin, Veteran, and Living Forward

The martial cultures explored in this book are, for the most part, life-long commitments. Within the Japanese *samurai* existed a name for those who no longer had either the ability or personage "to serve" and therefore are no longer *samurai*. These were known as *rōnin*, often translated as "masterless samurai" (Turnbull *The Samurai* 181). Caiger and Mason assert

> samurai frequently became rōnin ("wave men," i.e., vassals without a lord), drifted about, and sank into the ranks of commoners. In the seventeenth century, men often found they were rōnin through no fault or wish of their own, as a result of a major political upset such as the enforced transfer of a daimyo or confiscation of a domain [....] But an individual samurai was always free to sever relations with his ancestral lord voluntarily, and many took this step. (220–21)

Rōnin may have continued to hire out to other lords as swords for hire. However, it must be remembered that *samurai* were raised not just in ways of the warrior but were also educated:

> the most interesting and fruitful element in Tokugawa education was [...] the various private academies (*juku*). These schools specialized in a single branch of learning, or in the military arts, at a fairly advanced level. [....] Socially, *juku* indicate the existence of a semi-independent national intelligentsia after about 1750, one that had little to do with the conventions of class and regional differentiation. Intellectually, they embraced the whole range of available learning, and some concentrated exclusively on Western studies. [....] The *juku* are also interesting because many of the scholars who ran them were samurai, often rōnin. (Mason and Caiger 248–49)

Rōnin, free of the bureaucratic burdens of the samurai, were able to expand into "education and intellectual life" (Mason and Caiger 249) and sought to improve themselves and others not just through the *juku* but also through other arts.

Chikamatsu Monzaemon, "the great [*kabuki*] dramatist" (Mason and Caiger 234), is thought to be a *rōnin* (Mason and Caiger 237). At the core of "Chikamatsu's works is [the conflict] between a sense of duty (*giri*) and human feeling (*ninjō*) [wherein] the persons concerned satisfactorily reconcile their emotions with their duties to society." (Mason and Caiger 235). Another *rōnin*, Matsuo Bashō, who was "born into the samurai class, but chose to leave it when he grew up and make his way as a commoner"(Mason and Caiger 237) invented the haiku poem. "'Bashō' became virtually a household word while he was still alive and attracted patrons, associates, and pupils from all classes and both sexes" (Mason and Caiger 237). The *rōnin* who taught at the *juku*, Monzaemon, and Bashō all serve to demonstrate something unique in the world of martial cultures. Martial experts moved into other areas, making contributions to intellectual and artistic expansion for not just martial culture but the wider society and eventually the world.

Further, Monzaemon's concept of *girininjō*, torn between communal duty and personal agency, may find natural resonance with adherents to US martial culture which finds itself out of balance (Mason and Caiger 236). If martial cultures do not allow for individual freedom on some level servicemembers become trapped between conflicting wants and desires. In many cases, such servicemembers will come to a point where they feel the pull between a need to answer a call to duty or to explore areas of life that may be considered in conflict with codified morality-based laws that may or may not be necessary or healthy for the practitioners within the martial culture.

The Fuke sect of Zen Buddhism consisted of *komusō*, roughly translated as priest (or monk) of nothingness. They are identified by their *tengai*, a unique wicker hat, and a singular instrument known as *shakuhachi*, a long bamboo flute. This sect "only allowed in men of the samurai or ronin class" and played *honkyoku* (compositions) as a way of meditating, focusing "their minds toward enlightenment; they called it *suizen*, or 'blowing zen'" (Grundhauser).

Many veterans find that they move forward into the world of civilian culture wanting to retreat into anonymity, to begin anew, like the ronin-turned-komusō. And in so doing, they can bring their own music and peace to the larger society that encompasses both martial and civilian cultures. Yet this can also be damaging, if they attempt to suppress, even in the privacy of their own selves, their experiences, good and bad, in service. For even in the Fuke sect, the very instrument of peace was also utilized as a weapon of self-defense (Ribble 7).

The *Book of Five Rings*, written by famed samurai Miyamoto Musashi, was intended as a guide to future spiritual inheritors of martial culture. Musashi teaches "In all forms of strategy, it is necessary to maintain the combat stance in everyday life and to make your everyday stance your combat stance" (Miyamoto and Harris 54). Musashi's dictum implies that whether one is a servicemember or veteran one should accept the life-long existence of martial culture and not be afraid to extend not just their external stance (or bearing) but also their inner selves as committed to serving others, as the words *samurai* and servicemember represent.

Today

Amy Schafer's findings regarding the increase in blood-tied families who serve in the military led her to assert that this data may signal the creation of a US military "warrior caste" (Schafer). One of the popular, prominent examples of warrior castes is the *samurai*. However, the examination of Japanese martial culture and the mythology that underpinned their existence in this chapter indicates that there is far more nuance and misunderstanding when generalizing about martial cultures in this way. Further, the diversity within martial cultures, such as the *samurai*, demonstrates how quite different an interpretation of a martial culture can become once it is examined.

The *Kojiki* has heavily influenced older Japanese martial cultures. The sanctity of the land as *kami*, the ubiquity of *jinen* as an all-encompassing Nature, the importance of purification, and the necessity of service are all paradigms that US servicemembers today can learn from and apply to their lifeway today. There must be a balance rather than striving for a false perfection that requires casting aside "beautiful sorrow." Veterans and servicemembers have much to learn from the mythic *Kojiki*, the multitude of martial traditions, and the blending of the old ways (Shintō) with the new (Buddhism) to embrace an experiential life-affirming approach in Zen. Further, the examples set by *rōnin* of the past speak to current veterans of all nations that they have within them an energetic vitality and further gifts to give to the world.

Rōnin established both a precedent and example for veterans today in their example of establishing roles in education, intellectual, and artistic pursuits. The *shirasaya* is an outward physical symbol of the transformation. If the sword blade is the soul of the warrior, the fact that the blade can be "dressed" for both war and peace reflects the state of one in martial culture. When active in the wilderness beyond, one adopts the *koshirae* dressing (handle, wrap, pommel, handguard, etc.). After which the blade is placed in *shirasaya* fittings. These protect the blade from oxidizing by creating an airtight seal, and in appearance, they portray a beautiful smooth polished piece of wood with no apparent break. Over the years, the wood can become scratched, pitted, rough, and perhaps covered in dust, unrecognizable from its initial polished form, even to itself. Yet within dwells the protected warrior soul—ever-ready—at peace. If veterans are to move forward and grow, they must incorporate their whole selves bringing their art, music, and life fully into the world while maintaining the inner servicemember as the *shirasaya* conceals the blade; blooming as cherry blossoms under the dawning light of Amaterasu.

Notes

1 Buddhism was introduced to Japan in the fifth century CE. Buddhism came to the forefront when the Japanese Yamato court was introduced to Buddhism "sometime during the sixth century A.D." (Kitagawa 155). Yet it was the varieties of Mahayana Buddhism that really took hold in Japan. "As a missionizing religion, Buddhism grew rapidly in Japan. The immensely popular Mahayana Buddhism sects Tendai and Shingon eagerly sought to absorb Shintō. In Heian times, Shintō shrines throughout the country were taken over by Buddhist priests. The deities for whom the shrines had originally been built were now esteemed as minor manifestations of the cosmic Buddha [....] This amalgamation of Buddhism and Shintō (called Ryōbu Shintō, or Twofold Shintō) was the dominant form of religion in Japan from the eleventh century to the mid-nineteenth century. (Mason and Caiger 108)

2 Confucianism was introduced into Japan in the sixth century CE by Prince Shook and "seems to have been the first Japanese to proclaim principles intended to

support properly a centralized state under imperial rule" (Mason and Caiger 41) which mythologically descended from the Sun Goddess Amaterasu's grandson Jimmu, descended to Japan to rule the earth as the first emperor and the bloodline is viewed as an unbroken string of divinity or kami influence on earth. The "word for "government" in ancient Japan was matsurigoto, meaning "the business of worship." (Mason and Caiger 32). The sacrosanctity of the imperial line in combination with the presence of numerous kami shrines clearly indicates Japan is infused with divine power creating a world conscious of the numinous in all things. To devote oneself to the protection of the land or preservation of the people is to enter into a sacred act. Any defense of Japan is in service to a sacred mythological phenomenon. Just as in previous chapters, the sacrosanctity of the land and the devotion to serve as a protector of that land through utilizing force is a common theme and is reflected in US martial culture.

3 While Chamberlain notes that he understands this to mean they died, both his and other translations imply not a death but a concealing of themselves within the universe (Chamberlain and Aston 17; Yasumaro 7).

References

Bender, Ross. "The Hachiman Cult and the Dōkyō Incident." *Monumenta Nipponica*, vol. 34, no. 2, Sophia University, 1979, pp. 125–53, https://doi.org/10.2307/2384320

Breen, John, and Mark Teeuwen. *A New History of Shintō*. Wiley-Blackwell, 2010.

Chamberlain, Basil Hall, and William George Aston. *The Kojiki: Records of Ancient Matters*. Tuttle Publishing, 2012.

"Chosin Reservoir." *Naval History and Heritage Command*, US Navy, https://www.history.navy.mil/content/history/nhhc/browse-by-topic/wars-conflicts-and-operations/korean-war/korea-operations/chosin-reservoir.html

Cleary, Thomas. *Soul of the Samurai: Modern Translations of Three Classic Works of Zen & Bushido*. Tuttle Publishing, 2005.

———. *Training the Samurai Mind: A Bushido Sourcebook*. Shambhala, 2009.

Ebel, Jonathan H. *G.I. Messiahs: Soldiering, War, and American Civil Religion*. Yale University Press, 2016.

Fusé, Toyomasa. "Suicide and Culture in Japan: A Study of Seppuku as an Institutionalized Form of Suicide." *Social Psychiatry*, vol. 15, 1980, pp. 57–63, https://doi.org/10.1007/BF00578069

Gibbons-Neff, Thomas. "A Veteran Photo Project That Shows What Can't Always Be Spoken." *The Washington Post*, WP Company, 5 Dec. 2014, https://www.washingtonpost.com/news/checkpoint/wp/2014/12/05/a-veteran-photo-project-that-shows-what-cant-always-be-spoken/

Glassman, Hank. *The Face of Jizō: Image and Cult in Medieval Japanese Buddhism*. University of Hawai'i Press, 2012.

Gonzales, Amy. "Hurlburt Commemorates Desert One 26th Anniversary." *Hurlburt Field*, Air Force, 8 May 2006, https://www.hurlburt.af.mil/News/Article-Display/Article/206148/hurlburt-commemorates-desert-one-26th-anniversary/

Grundhauser, Eric. "The Bamboo Flutes of Japan's 'Monks of Emptiness.'" *Atlas Obscura*, 10 June 2021, www.atlasobscura.com/articles/komuso-flute-monk-japan-basket-head-zen-buddhism.

Haldén, Peter, and Peter Jackson, editors. *Transforming Warriors the Ritual Organization of Military Force*. Routledge, 2016.

Kawai, Hayao. *Dreams, Myths & Fairy Tales in Japan*. Edited by James Gerald Donat, Daimon, 1995.

Keeler, Matthew. "'The Hell We Went through': Korean War Vets Who Fought in Bloody Chosin Battle Honored in Seoul." *Stars and Stripes*, 29 Sept. 2019, https://www.stripes.com/theaters/asia_pacific/the-hell-we-went-through-korean-war-vets-who-fought-in-bloody-chosin-battle-honored-in-seoul-1.601021

King, Winston L. *Zen & The Way of the Sword: Arming the Samurai Psyche*. Oxford Univ. Press, 1993.

Kitagawa, Joseph M. *On Understanding Japanese Religion*. Princeton University Press, 1987.

Maiese, Aniello, et al. "A Peculiar Case of Suicide Enacted Through the Ancient Japanese Ritual of *Jigai*" *The American Journal of Forensic Medicine and Pathology*, vol. 35, no. 1, March 2014, pp. 8–10, https://doi.org/10.1097/PAF.0000000000000070

Mason, Richard H., and J. G. Caiger. *A History of Japan*. Revised Ed., Tuttle, 1997.

McCoy, Brady. Personal Interview. 6 June 2023.

Miyamoto, Musashi, and Victor Harris. *The Book of Five Rings*. Sirius, 2021.

Ono, Sokyo and William Woodard. *Shinto: The Kami Way*. Tuttle Publishing, 1962.

Osuna, Freddy. Personal Interview. 14 June 2023.

Ravina, Mark. *The Last Samurai: The Life and Battles of Saigo Takamori*. John Wiley & Sons, Inc., 2011.

Ribble, Daniel B. "The Shakuhachi and the Ney: A Comparison of Two Flutes from the Far Reaches of Asia." *Bulletin of Kochi Medical University No. 19*, 2003, www.shakuhachi.com/K-Ribble-ShakuhachiandNey.pdf

Sadakata, Akira, and Hajime Nakamura. *Buddhist Cosmology: Philosophy and Origins*. Kōsei, 1997.

Schafer, Amy. "Generations of War: The Rise of the Warrior Caste and the All-Volunteer Force." *Center for a New American Security (En-US)*, 2017, www.cnas.org/publications/reports/generations-of-war

Stone, Jacqueline Ilyse, and Mariko Namba Walter. *Death and the Afterlife in Japanese Buddhism*. University of Hawai'i Press, 2008.

Suzuki, Daisetz Teitaro, and Richard M. Jaffe. *Zen and Japanese Culture*. Princeton University Press, 2019.

Turnbull, Stephen. *The Samurai: A Military History*. Routledge, 2007.

Turnbull, Stephen R., and Giuseppe Rava. *Samurai Women, 1184–1877*. Osprey Publishing, 2010.

Tyler, Royall. *The Tale of the Heike*. Kindle ed., Penguin Publishing Group, 2014.

Yamamoto, Tsunetomo. *Hagakure: The Secret Wisdom of the Samurai*. Translated by Alexander Bennett, Tuttle, 2014.

"2021 National Veteran Suicide Prevention Annual Report." *U.S. Department of Veteran's Affairs - Mental Health*, Sept. 2021, https://www.mentalhealth.va.gov/docs/data-sheets/2021/2021-National-Veteran-Suicide-Prevention-Annual-Report-FINAL-9-8-21.pdf

7   Conclusion

As I was finishing the manuscript for this book, I attended two events that showed the width of the gap between the outsiders who seek to "fix" veterans and the holistic, lived experience of those in martial culture. I attended a presentation in July of 2023 in which a humanities professor of classics presented her work with veterans in providing small-group reading and reflection sessions. She used Roman and Greek sources—primarily the Odyssey and Iliad—as well as other poetry. She referred to her two favorite poets: one who was a spouse of a servicemember and another who was a veteran of the Vietnam War. Readings from these poets, which dominated much of the presentation, were completely focused on intense pain, loss, and anger. She discussed focusing almost completely on Book 8 of the Odyssey to get the "really good" conversations. What she meant by better seemed to focus on emphasizing cognitive dissonance, pain, and pathology, and her bias was apparent.

It was clear that despite facilitating these groups for years, she still did not have the correct language for martial culture. This professor is a guest speaker on veteran work in many places, and the Veterans Administration has hosted her to speak with veterans. Her presentation was part of a series on how to engage humanities studies in veteran mental health care, and she spoke with confident authority despite her lack of education in martial culture or mental health treatment. She is a Classics professor providing mental health observations about a culture she has only minute, biased, and second-hand knowledge of.

In her introduction, she stated that we are currently in the post-Vietnam era. As a veteran who has taught and learned military history within professional military education environments, I was stunned. It has been 48 years since the end of the US involvement in the first televised war, which is 18 years longer than the time between the end of World War II and the beginning of the Vietnam war. So, for someone to say that we are in the post-Vietnam era would be the equivalent of someone referring to 1993, two years after Desert Shield/Desert Storm, and the same year as Operation Gothic Serpent in Somalia (more commonly known by the film Black Hawk Down) as the post-World War II era.

DOI: 10.4324/9781032613222-7

The characterization reveals the speaker's disorientation in martial culture, which seems common. Each civilian generation defines martial culture by their personal biases and experience, whether through various readings or media consumption. And as generations age, finding themselves in positions of authority in public spaces, they seem to project their understanding and their personal biases regarding conflict and what they believe US martial culture is, onto the following generations.

Of course, that viewpoint is in error because martial culture constantly evolves and changes as all cultures do. It should be noted that World War II, Korean, and Vietnam veterans all served in a martial culture which was a mix of volunteers and conscripts personnel. That alone should illuminate a cascade of differences. But of course, it's not the only difference. That is not to say there are no similarities between martial cultures. However, the way in which Vietnam colors any discussion or perception of events in the United States is overemphasized and, I believe, detrimental.

Two days later, I attended a very different event. I attended a friend's retirement from the Air Force and so doing, witnessed a true celebration of a martial life. People I had not seen in years came to this event, where there were hugs, huge smiles, tears of joy, and an immediate synchronization of intent and oneness as we came to celebrate this person's life.

When I met the retiree years earlier, we were both NCOs serving in the same unit. He was now retiring as a Senior Master Sergeant. To reunite with him and others again recalled deep emotions. Years melted away in an instant, and the mix of retirees and servicemembers, from general officer on down, became fellow celebrants.

The ceremony was a chain of beautiful individual rituals. They gave a sense of renewed belonging through services. The rites invoked the shared experiences of thresholds past and trials overcome renewing the recognition and acknowledgement of the reality of those experiences.

One of the most immersive rituals was the telling of the servicemember's life. From childhood to that day, the retiree's story was told by our former commander. This elder in dress uniform spoke the story of my friend's life—the triumphs and tragedies, the humorous, and the heart-wrenching. It was a celebration of a martial lifeway brought to the present day. People from the audience who had served with him in various units were called to stand with the servicemember, in recognition of the necessity of interconnectedness. There was a calling forward of all who had earned the beret. All of those who had, some in uniform still serving, many not, stepped forward.

The retiree was presented with gifts of symbolic power. Among these were cloth flags adorned with unit symbols, a wooden hammer shaped like Mjolnir (Thor's hammer), a Roman rudis (wooden gladius sword) as a symbol of freedom, as well as the American flag, described as holding a constellation of stars. He presented gifts to each member of his family. He read texts aloud regarding the sacrifices that they had to go through in their unique place within martial culture. He talked again about his heritage and martial culture. He concluded that one must become a reader; to learn to love to read and grow from the reading. And lastly, tell people you love them. He asked the audience to not let arguments stop one's conveyance of love. He recalls the father he lost, the airmen he lost. Through tears he invokes their names, and in the silence that follows the ceremony, and final ritual, is complete. Two very different experiences, outsider and insider, of relationships with martial culture separated by such a small amount of time.

In the introduction to *The Bhagavad Gita: A Walkthrough for Westerners*, Jack Hawley asserts that we "have to keep reminding ourselves that the battle is metaphoric, that the war is being fought inside each of us and will continue throughout our lives" (xxxiv). For those in US martial cultures today, this statement is both true and false: to exist in martial culture struggle that is both internal and external. The battles can be very real and the myths explored in this book are relived every day. As this book is written and read, a recruit is paired with another to form a "battle buddy team" in basic training. They are twinned—as the two Navajo brothers were—and tested by their instructors, just as the gods assess the Navajo brothers. And both the modern US recruits and the Navajo brothers are found wanting. The recruits will be evaluated again, just as the Sun tests the two Navajo brothers, and finally allowed to pass through the threshold from recruit to full member of the military organization and martial culture.

Servicemembers are training today, as the Pandava brothers did under Bhishma, Kripa, and Drona. Elsewhere, seasoned servicemembers step off the ramp of a plane, setting foot on land to conduct their mission, just as the two Navajo brothers alight on Turquoise Mountain, or the Pandavas drive their chariots onto the plain of Kurukshetra. These seasoned servicemembers may imitate great natural forces, summoning within them a likeness of a bear, wolf, or thunderstorm just as Norse berserkers did. Or perhaps, because those natural models of unrestrained power are no longer as present in the lives of today's peoples, the servicemembers may use as their examples movie characters or superheroes that popular culture has imbued with supernatural forces such as the Wolverine, John Rambo, the Batman, The Punisher, or Captain America. Others sit in the dark, contemplating their blessings and curses, as Starkad was blessed by Odin and cursed by Thor; they are arrayed with glory but mourn the loss of hearth and family.

While veterans, weary of service and worn by trials kneel by the river as Izanagi does, to purify themselves and find their place in a world, like wandering *rōnin* seeking reconnection with *jinen*, all-encompassing Nature. Other veterans seek connection and safety through their animal companions, much like Yudhishthira, who refuses both Indra and heaven and chooses to not abandon his companion dog. And lastly, the servicemember or veteran who passes from this life may journey to the halls of Freya or Odin.

The living *mythos* of martial culture happens every day around the world. The battle on the plain of Kurukshetra in the *Mahabharata* takes place in Mosul, Iraq, the Korengal Valley of Afghanistan, and Syria. US servicemembers unknowingly reenact Yudhishthira's emotional paralysis, sitting in the dust outside the gate of the city, as they sit in their cars outside their homes, breathing through anxiety. Servicemembers and veterans look for their Bhishma, their Holy People, their Thor, their purification by the river after having escaped the underworld as the *kami* Izanagi does. The myths are alive, and yet servicemembers and veterans are largely unaware of their participation in them, reincarnating mythemes with every turn of the earth.

Seeking *mythos* over *logos* was recently encapsulated in an exchange between a young airman and a general officer. The airman asks this question (paraphrased from the original): "General…do you have any advice on having a successful career vs a significant career? A successful career could be seen as you made General, but a significant career could mean you made an impact on lives and left it on all the table" (McCoy). This question belies a hunger for—not a career—but a lifeway. Yet this airman only has the words available (career) because of the secular bureaucratic nature of the military organization as part of a government organization. The re-introduction of the myths, not as quaint fables or fictions but a true mythical and spiritual inheritance from the ancients, will revivify martial culture and answer that young servicemember's need for a lifeway of value rather than a secular "successful career."

The question speaks to what is missing: the re-souling of the military organization by the recognition of martial culture as an existing phenomenon. US martial culture must be wholly reconnected to the ancient understanding of life-long engagement in service. This is a type of re-membering, putting the culture back together through re-joining it to the songs of the elders in the form of the mythologies examined here and the many, many more that have yet to be explored through this lens. Key is the recognition that servicemembers who become veterans will eventually look into Amaterasu's mirror to reflect on who they are—all that they are—and find completion and wholeness.

This book defines and describes martial culture as a separate and unique lifeway, conducts a hermeneutical study of select mythologies from disparate regions and times—combined with contemporary qualitative experiences of US servicemembers who have transcended the military organization—to explore correlations between the two and find what in the selected myths addresses the challenges of US martial culture today. A renewed connection to ancient forebearers in other martial cultures and recognition of their "spiritual inheritance" through that connection will aid in balancing servicemembers and veterans and provide a foundation for and a language to express the transcendental experiences of servicemembers today.

US martial culture exists as secular, ancestrally oriented, nomadic, totemic, chiefdoms of tribal social groups. As described in this book, US martial cultures maintain their own historic and contemporary personalities as guides to behavior and are heavily imbued with symbolism. The individual exists to support the whole and gains self-worth from the whole, with loss of membership to the active military tribe resulting in the perceived loss of self-worth. Martial cultures are hierarchal with internal caste-like rules of association where every member of the society is a potential regulator of behavior. Regulation and inclusion extend to the immediate family of the servicemember, who also exist within martial culture. Nomadic, at least in the active duty military, the servicemember and their immediate family exist in a fluid liminality.

Older martial cultures had mythologies that sustained their warriors and provided the essential songs, dances, and "psychological center" (Campbell and Moyers *Ep. 1*). The wisdom and knowledge of how to live a martial life of service has been lost to corporate, secular approaches to constructing and operating military organizations that emphasize *logos* over *mythos*. Insistence on industrialized approaches to service has led to an atrophying of the rich transcendent-touching experiences that are experienced every day, because the adherents to martial culture have no words and no songs to describe them.

While attempts to reconnect with the soulful aspects of martial culture are occurring in small ways, such as retreats organized by veterans for veterans, servicemembers, and their families, a true reset is needed. The military organizational emphasis on career over lifeway and failure to acknowledge the rituals and myths by referring to them as customs and traditions (to make more palatable to a secular government) are just a few of the industrialized approaches that reduce the wholeness of martial culture. There is an acute need for a holistic reformation effort recognized and supported by both the informal and formal institutions of the military and government. This remythologizing should be done in a life-affirming way, yet also demonstrate the *gravitas* of choosing to volunteer to enter a martial culture. The message of the myths will contradict the current marketing and false narratives regarding the temporary nature of service and that one can enter and leave military service without being changed into something else.

Perhaps many parents will not want to openly acknowledge their children (now young adults) will be choosing a different culture. Young adults may hesitate in the face of the reality that their mannerisms, values, beliefs, and norms will be altered by choosing martial culture. To choose military service is to step through the looking glass. Further, a recognition of martial culture will dispel the lies told at the other end, with phrases like "I can't wait to go back to being a civilian again." The danger of believing these false narratives is evident in an interview with a 50-year-old retired chemistry professor with no military experience who attempted to join the Ukrainian foreign legion in its resistance of the current Russian invasion. "If they have room for one more, I'm here, and I'm ready to sign up and go," however he was rejected because he would not "sign a contract that would have bound him until the conflict ended. [....] 'I can give you a month. But I have a job and family to get back to. I have a life.'" (McDonnell) This individual, as a person of civilian culture, believes one can place their toe in the water without getting wet. That is the implication of the idioms "go back to being a civilian" and "get back to the real world." The true gravity of the commitment to step through a looking glass is missing. This, ironically, is still how most servicemembers view their situation, not realizing the depths of martial culture.

Full recognition and awareness of martial culture will strip away comfortable illusions to reveal the reality of the expanse of martial culture, much to the possible angst of the potential recruits, recruiters, politicians, and even current servicemembers. Placing one's foot on the path, as the two *Diné* brothers did to find the Sun, is to begin a metamorphosis that will be unpredictable in its outcome. J.R.R. Tolkien, veteran of World War I, captured in his book *The Hobbit*. In a film adaptation, the sentiment is implicitly expressed when Bilbo asks, "Can you promise that I will come back?" to which Gandalf responds, "No. And if you do, you will not be the same" (Jackson). Even so, to be on the pollen path of the *Diné* is to begin a sacred journey, and if one is truly ready for it, they will find who they truly are. For while, the servicemember and veteran may be a different creature than they were before, that does not mean that the change should not have occurred. In many interviews with veterans, they do not regret what came of their choice to enter US martial culture and affirm they would do it again.

Acknowledging martial culture exists will allow for greater accuracy in identifying the difficulties veterans face such as blanket diagnoses of PTSD, alcohol and drug abuse, isolation, and suicide. Further, the myths, as demonstrated, provide examples for how servicemembers and veterans can and should provide for their own well-being, which in turn will serve the world. As Campbell once stated, in saving oneself "you save the world [....] any world is a living world if it's alive, and the thing is to bring it to life. And the way to bring it to life is to find in your own case where your life is, and be alive yourself" (Campbell and Moyers *Ep. 1*). The veteran must stay in touch with the martial culture she/he comes from and, like the *rōnin*, bring the best of that culture—combined with their own unique gifts—with them wherever they may go.

Martial culture theory attempts to reintroduce and reintegrate the multidimensionality of human existence within and beyond military service through a useful revivification of mythology. Martial culture exists as separate from and, in many ways, is largely defined in relation to civilian cultures. Martial and civilian cultures may find ways to emphasize differences; however, over-emphasizing or exaggerating traits found negative by the "other" interferes with bridging the gap in civil-military discourse. Neither veterans nor civilians should feel empowered over the other or cultivate an elitism in either direction. Today these cultures are in a symbiotic relationship; each relies on the other and is bound to each other.

Warrior versus Servicemember

Calls by veterans and servicemembers asking to cease calling modern US servicemembers warriors—or worse still, heroes—demonstrate the resonance with the recurring theme of service. The revivification of the protector form of Mars—Mars Quirinus—brings a holistic balance to the martial. The renewal of understanding that the shorthand "translation" of *samurai* to warrior, in fact, misrepresents the term, which is more closely related to servicemember. Service is the common root from which these martial cultures spring. A reconnection with mythology, especially those myths outside of the Greek and Roman worlds, would bring in a much needed reemphasis of service.

Communitas

The mythologies explored in this book demonstrate aspects of familial or tribal nature. The necessity of elders is essential to this form of culture and necessary for familial inclusion. The mythologies speak to the closeness of those engaged in service who are largely bound through actual familial connections. US martial cultures demonstrate how those bonds can extend past blood relations, to form communities and bonds akin to or even stronger than blood relationships.

Gender

The findings of this book show that the role of genders other than male within martial cultures has been suppressed. Gender-role norming has created false concepts manifested in lists of masculine and feminine traits. In a study concerning competitiveness in women, Alessandra Cassar and Mary Rigdon note "women are not less competitive than men" if the competition has a prosocial aspect (Cassar and Rigdon). Considering that martial culture has an inherently prosocial aspect, combined with the already-noted discoveries of women as far more active in combat than history has acknowledged, leaving little doubt that many of the narratives passed down through history regarding the feminine and warrior are being disproven today.

The recent decision by Ukraine, as of this writing currently under invasion from Russia and facing existential threat, to ban the fleeing of any Ukrainian male reinforces the false narrative—as demonstrated throughout this book—that "war is the province of men" (Maguire). It comes as no surprise to those familiar with the millennia of active engagement by women in conflict that many Ukrainian women have volunteered to fight, whereas some men are currently being pressed into service (Bloom and Moskalenko). And as argued in chapter two, familial connections to martial culture are powerful motivators. Tany Kobzar, a Ukrainian 49-year-old and mother, enlisted in the army after the 2020 invasion of Ukraine by Russia. As the news filled with threats of an impending Russian invasion, she was haunted,

> "I was waking up in the middle of the night, terrified. I would look at a black-and-white photo of my grandmother" [....] she recalls. "She reminds me of how brave a person can be." Kobzar's late grandmother was an army medic in World War II. (Frayer and Matviyishyn)

Kobzar and her grandmother join the ranks of the *Onna-Bugeisha* or *Onna-musha* (female *samurai*), shield maidens, and others detailed in this book. Archeological, historical, and anthropological studies demonstrate women have always been part of martial cultures. While in many cases both women and men have, throughout human existence, not been given a choice about whether to participate in or experience war, many volunteered to serve when it was never expected of them. It is my hope that the Ukrainian people will not suppress the recognition of the active role of women in combat as has been the norm throughout the world.

Land

The sacred is a recurring theme throughout this book. To be sacred is to be set apart from the world. The sacredness lies not just in a totem such as a symbolic patch but in the land, as described in the encircled lands within the four Holy Mountains of the *Diné*, or the sacred groves dedicated to Thor, or the forests and waters of Japan where *kami* dwell. And thereafter in past martial cultures, to serve that land, the *jinen* (Nature) was to participate in a sacred act.

Because being born in a land is recognized as being native to that *jinen*, some Indigenous peoples increasingly prefer to be referred to as First Peoples, First Nations, or Indigenous Peoples. Yet to a considerable extent, the lands that constitute the United States are not recognized as inherently sacred. Many servicemembers and veterans have come to realize the importance of "home" in the need to reconnect with *jinen* through hiking, fishing, spending time walking the streets of one's city or town, or even standing in a cleansing rainfall. There needs to be a larger, official recognition of the land as alive and servicemembers as serving not only the protection of the peoples but the mountains, skies, rivers, lakes, plains, plants, animals, and even the sacredness of cities. Such a recognition would help strip away any tendencies to serve a particular leader, political agenda, or ideology and return a sense of balance and humility before *jinen* to acts of service.

Threshold Experiences

The myths explored in this book illustrate one is not simply born as Athena, fully grown with armor and immediately capable with all knowledge and skill ready to move forward into battle. Basic training is a culling process whereby the individual is shaped and molded. A "becoming" must take place, and many of the trials are designed for only those who have succeeded in earlier challenges. The five Pandava brothers, even their sixth brother Karna, who was born with armor, required teachers and years of training to unlock their innate abilities and temper useless attitudes and practices.

The necessity of training and testing is demonstrated in both the Navajo myth of the two brothers, also born of celestial deities, who nonetheless play with bows and arrows made by their mother as children before being taught and tested by gods. Recall the first visit when they fail to outrace the gods. Sometimes the individual fails the first test and then must try again in a year, or two, or even longer.

Servicemembers as Sacred Warriors

Many of the mythologies explored in this book speak to a necessity for balance between the warrior and the holy person, as in the examples within the *Mahabharata* where the union between *kshatriya* and *brahmin* ushered in a Golden Age. The recurrence of warrior/priest figures illustrates the sacrosanctity that must be respected in living a martial lifeway. The *samurai* relationship with Zen Buddhism and Shinto is another illustration of using force when necessity demands without losing sight of sanctity, not only of the sacredness of one's own but also seeing the sacred within the "other" with whom the servicemember is in conflict.

Restoring and reinvigorating the mythologies does not imply a need for a specific religious doctrine but an understanding that the service itself is not a vocation, but a sacred act, connected to *jinen* (Nature) and born of compassion combined with a type of stoic understanding of duty which in this book has been encountered throughout in terms of *dharma* or *Bushido*.

Cyclic Existence and Purification

Each mythology in this book demonstrates what a life-long existence in a martial culture means in terms of cyclic patterns associated with going forth and returning from the field or wilderness, the places beyond the gates that represent the barrier between the home and the wildness. In some cases, this barrier is physically marked, whereas in others it must be consciously recognized, and in many cases passing from one to the next requires an active participation in a ritual.

This book has revealed that in all four mythologies there is a need for cleansing rituals in martial cultures. "Yet even here the warrior's situation is ambiguous, for, having tasted blood—even in a good cause—he is a potential danger to the peace and well-being of the rest of his society and must typically undergo a rite of purification before being readmitted to it." (Dumézil *Gods of the Ancient Northmen* xiv). The Navajo brothers perform the Enemy Way Ceremony upon returning from the wilds and Izanagi kneels beside the river to wash.

Veteran and Moving Forward

The term *veteran* was explored as a recent invention with the rise of civilian life-ways that have, in our time, eclipsed martial cultures in size and influence. For warriors of First Nations such as Kiowa and the Chiricahua Apache, martial culture is a life-long commitment (Meadows 3–7; Mort 120–21). Further, the very need to have a term such as veteran demonstrates a recognition that martial culture exists beyond membership in a military organization, much in the same way *rōnin* proves a similar recognition in discussions about Japanese martial cultures.

This ability to move forward, as shown in the *Mahabharata*, is both necessary and incredibly difficult. Yudhishthira's crisis after the battle when he experiences doubts of living a kshatriya's life after having fought and won against those who he loves is an example of the same crises experienced by US servicemembers and veterans today. In many cases, they are wracked by the inability to move forward and also vow to "drift like the wind about the world until the dissolution of this body" (Satyamurti 657). Yet Yudhishthira is counseled to go to Bhishma, who guides him, reminding Yudhishthira that he is still a *kshatriya* and the world needs him to move forward in service, Yudhishthira is saved. Many US service-members and veterans lack a Bhishma, though some find one who can embody the proper guide through veteran groups. These elders of martial culture must increase their roles in aiding those who come after, rather than relying on secular government institutions to provide re-souling answers.

External Aid

Two oft-heard phrases in today's US military is "Take care of your family because they are all that will be left once you are done with the service (or the service is done with you)" and the second "You need to do these things in order to get promoted or advance in your career," implying one's inclusion in the military organization hinges on adopting careerism. Both are true in a secular bureaucracy that determines the servicemember is expendable hardware because the servicemember is considered a financial resource to either be invested in, wrongly supported to "protect the investment," or abandoned when no longer necessary. However, if a fully functioning martial culture is nurtured, the above statements will become false. The "expendable" nature inherent in industrial secularist views demonstrates the cliff that is artificially created in the refusal to affirm that martial culture extends well beyond placing a uniform in storage. Refusal to acknowledge martial culture pushes a family unit, not only the service-member but the spouse and children, out into civilian culture without realizing why they might experience loss. Acknowledging martial culture will allow them to maintain and cultivate those relationships within the martial sphere, reducing culture shock and preventing dis-ease.

The mythologies of martial cultures possess keys to doors servicemembers do not even understand exist. Hayao Kawai discusses differences in Western views of perfection in fairy tale endings in terms of Western "happy marriages" versus Japanese "sorrowful beauty" in which both are balanced from the views of the respective cultures (Kawai 122). Cultural differences can be wonderful for discovery, interaction, and friendship. However, Kawai's work suggests that one cannot easily work cross-culturally to provide psychological assistance. Without understanding and acceptance of the symbols, language, values, beliefs, and norms of the individual's culture, one risks polluting any sort of assistance through cultural contamination or cultural value projection.

The culture must create its own healers and guides as demonstrated through the examples in the myths. Having witnessed a discussion between a Vietnam veteran and veterans of Iraq and Afghanistan, the transformations that occur in the inter-generational exchange in terms of helping heal one another are great. This role of elders is key to demonstrating that there is a future past the present torment, and that there is no reason to withhold oneself from the world. The concept of "healing through the crowd," as in the example of the two Navajo brothers healed by the Holy People, has power that needs to be inculcated.

Reconnecting with martial cultures' spiritual ancestors through myth and story may also correct beliefs of being owed something for becoming a part of martial culture. That viewpoint is a corruption of service. Recalling the meaning of the word *samurai* or the *kshatriya* creed which emphasizes serving life, it must be understood that nowhere in the myths examined here is a pleasant life devoid of conflict guaranteed. If anything, to serve is to preserve those who need protection and liberate those who are oppressed while rediscovering a living ritual love of *jinen* (Nature). To serve is to take on a lifeway, not enter into a bartering agreement in which one is placed on a pedestal and granted comfort. Service is about experiencing life.

Most current members of US martial culture, servicemembers and veterans, do not have the connection with or understanding of *mythos* or their cultural heritage. The following tweet asserts that the current Chairman of the Joint Chiefs of Staff, General Mark Milley, "responds to a question about a shortage of behavioral health staff by saying that there is a 'behavioral health specialist' in every unit, and 'that's the squad leader'" (Myers). Of course, the person who relayed this via Twitter rightfully argues that while squad leaders are vital, they are not trained behavioral health specialists. Yet holding advanced degrees in behavioral health fields does not necessarily ensure one is the most appropriate person to treat servicemembers, veterans, or their families.

The veteran community must begin to provide for their own by getting those same degrees and certifications not only to treat those in martial culture but to bring martial culture into programs to enrich academic spaces and civilian culture. And further, the military organization must acknowledge the mythos of martial culture and begin to incorporate, from the very beginning, the knowledge of the sacred strands of relationship they share with their spiritual ancestors who served in martial cultures before them.

Conclusion

Joseph Campbell once wrote, "[W]hy should it be that whenever [humans] have looked for something solid on which to found their lives, they have chosen not the facts in which the world abounds, but the myths of an immemorial imagination?" (*Masks* 4). Through mythology and guided by those living elders who understand the mysteries and trials, servicemembers can reconnect with their ancient roots and learn from those who came before. A recognition of martial culture as a separate but valid lifeway can bridge the gap between those within and those outside to allow a place for martial cultures within the wider mix of diverse cross-cultural exchanges hoped for in the future of our species. If anything else, perhaps servicemembers and veterans will learn that to be a member of martial culture means one must learn the stories, dances, and songs of those who came before and that they, as a distinct cultural group, will be encouraged to dance their dances and sing their songs.

All of these mythologies tell a story of a continuous cycle of donning equipment and moving forward, returning home, and going out again... just as the Two Navajo Brothers live their lives. Out and back, out and back... the cyclic existence of a martial culture spins within the cosmic dance of the gods as we watch.

References

Bloom, Mia, and Sophia Moskalenko. "Ukraine's Women Fighters Reflect a Cultural Tradition of Feminist Independence." *The Conversation*, 6 Apr. 2022, https://theconversation.com/ukraines-women-fighters-reflect-a-cultural-tradition-of-feminist-independence-179529

Campbell, Joseph. *The Masks of God: Creative Mythology*. Penguin Books, 1976.

Campbell, Joseph, and Bill Moyers. *Ep. 1: Joseph Campbell and the Power of Myth "The Hero's Adventure."* 27 Aug. 2018, https://billmoyers.com/content/ep-1-joseph-campbell-and-the-power-of-myth-the-hero%E2%80%99s-adventure-audio/

Cassar, Alessandra, and Mary L. Rigdon. "Prosocial Option Increases Women's Entry into Competition." *Proceedings of the National Academy of Sciences*, vol. 118, no. 45, Nov 2021, e2111943118, https://doi.org/10.1073/pnas.2111943118, accessed 14 Sept. 2020.

Frayer, Lauren, and Iryna Matviyishyn. "Ukrainian Women Are Volunteering to Fight - and History Shows They Always Have." *NPR*, 19 Mar. 2022, https://www.npr.org/2022/03/19/1087712539/ukrainian-women-are-volunteering-to-fight-and-history-shows-they-always-have

Hawley, Jack. *The Bhagavad Gita: A Walkthrough for Westerners*. Reprint ed., New World Library, 2011.

Jackson, Peter, director. *The Hobbit, An Unexpected Journey*. Warner Bros., 2012.

Kawai, Hayao. *Dreams, Myths & Fairy Tales in Japan*. Edited by James Gerald Donat, Daimon, 1995.

Maguire, Amy. "Why Banning Men from Leaving Ukraine Violates Their Human Rights." *The Conversation*, 24 Mar. 2022, https://theconversation.com/why-banning-men-from-leaving-ukraine-violates-their-human-rights-178411

McCoy, Brady. "AETC CC's Order of the Sword Q&A." *Facebook*, 17 Apr. 2022, https://www.facebook.com/brady.mccoy.96/posts/pfbid0VdqgXWws3D5N9gX8u4bTG-krkv6tngEhbWCC2k88qBpyp6eQLMbkRKqDayPG9G65sl

McDonnell, Patrick. "U.S. Military Veterans Answer Zelensky's Call to Fight, But Not All Are Chosen." *Los Angeles Times*, 31 Mar. 2022, https://www.latimes.com/world-nation/story/2022-03-31/u-s-military-veterans-answer-zelenskys-call-to-fight-but-not-all-are-chosen

Meadows, William C. *Kiowa Military Societies: Ethnohistory and Ritual*. U of Oklahoma P, 2010.

Mort, Terry. *The Wrath of Cochise the Bascom Affair and the Origins of the Apache Wars*. Pegasus Books, 2015.

Myers, Meghann. "Gen. Milley Responds to a Question about a Shortage of Behavioral Health Staff by Saying That There Is a 'Behavioral Health Specialist' in Every Unit, and 'That's the Squad Leader.' Sigh." *Twitter*, 5 April 2022, https://twitter.com/meghann_mt/status/1511370872375894017?s=12&t=VOt8USuvbMhW__qKeB5GdQ.

Satyamurti, Carole. *Mahabharata: A Modern Retelling*. W.W. Norton & Company, 2015.

Index

For Product Safety Concerns and Information please contact our EU
representative GPSR@taylorandfrancis.com
Taylor & Francis Verlag GmbH, Kaufingerstraße 24, 80331 München, Germany